T0342377

Introduction to Mathematical Sociology

Introduction to Mathematical Sociology

Phillip Bonacich and Philip Lu

PRINCETON UNIVERSITY PRESS
PRINCETON & OXFORD

Copyright © 2012 by Princeton University Press
Published by Princeton University Press, 41 William Street,
Princeton, New Jersey 08540
In the United Kingdom: Princeton University Press,
6 Oxford Street, Woodstock, Oxfordshire OX20 1TW
press.princeton.edu

All Rights Reserved

Library of Congress Cataloging-in-Publication Data

Bonacich, Phillip, 1940-
Introduction to mathematical sociology / Phillip Bonacich and Philip Lu.
    p. cm.
Includes bibliographical references and index.
ISBN 978-0-691-14549-5 (hbk.)
1. Mathematical sociology.   I. Lu, Philip, 1982-   II. Title.
HM529.B66 2012
302.301′513–dc23
    2011041621

British Library Cataloging-in-Publication Data is available

This book has been composed in ITC Slimbach

Typeset by S R Nova Pvt Ltd, Bangalore, India

10 9 8 7 6 5 4 3 2 1

CONTENTS

List of Figures   ix
List of Tables   xiii
Preface   xv

1. Introduction   1
    Epidemics   2
    Residential Segregation   6
    Exercises   11

2. Set Theory and Mathematical Truth   12
    Boolean Algebra and Overlapping Groups   19
    Truth and Falsity in Mathematics   21
    Exercises   23

3. Probability: Pure and Applied   25
    Example: Gambling   28
    Two or More Events: Conditional Probabilities   29
    Two or More Events: Independence   30
    A Counting Rule: Permutations and Combinations   31
    The Binomial Distribution   32
    Exercises   36

4. Relations and Functions   38
    Symmetry   41
    Reflexivity   43
    Transitivity   44
    Weak Orders—Power and Hierarchy   45
    Equivalence Relations   46
    Structural Equivalence   47
    Transitive Closure: The Spread of Rumors and Diseases   49
    Exercises   51

5. Networks and Graphs   53
    Exercises   59

6. Weak Ties   61
    Bridges   61
    The Strength of Weak Ties   62
    Exercises   66

7. **Vectors and Matrices**   67
   Sociometric Matrices   69
   Probability Matrices   71
   The Matrix, Transposed   72
   Exercises   72

8. **Adding and Multiplying Matrices**   74
   Multiplication of Matrices   75
   Multiplication of Adjacency Matrices   77
   Locating Cliques   79
   Exercises   82

9. **Cliques and Other Groups**   84
   Blocks   86
   Exercises   87

10. **Centrality**   89
    Degree Centrality   93
    Graph Center   93
    Closeness Centrality   94
    Eigenvector Centrality   95
    Betweenness Centrality   96
    Centralization   99
    Exercises   101

11. **Small-World Networks**   102
    Short Network Distances   103
    Social Clustering   105
    The Small-World Network Model   111
    Exercises   116

12. **Scale-Free Networks**   117
    Power-Law Distribution   118
    Preferential Attachment   121
    Network Damage and Scale-Free Networks   129
    Disease Spread in Scale-Free Networks   134
    Exercises   136

13. **Balance Theory**   137
    Classic Balance Theory   137
    Structural Balance   145
    Exercises   148

14. **Markov Chains**   149
    Examples   149
    Powers of P, Paths in the Graphs, and Longer Intervals   154

The Markov Assumption: History Does Not Matter 156
Transition Matrices and Equilibrium 157
Exercises 158

15. Demography 161
Mortality 162
Life Expectancy 167
Fertility 171
Population Projection 173
Exercises 179

16. Evolutionary Game Theory 180
Iterated Prisoner's Dilemma 184
Evolutionary Stability 185
Exercises 188

17. Power and Cooperative Games 190
The Kernel 195
The Core 199
Exercises 200

18. Complexity and Chaos 202
Chaos 202
Complexity 206
Exercises 212

Afterword: "Resistance Is Futile" 213

Bibliography 217
Index 219

FIGURES

1.1 Total number of infected individuals when $pN = 1.5$ 3
1.2 Number of newly infected individuals over time when
    $pN = 1.5$ 4
1.3 Number of total and newly infected individuals over time
    when $pN = .75$ 4
1.4 Initial locations of actors in the Schelling simulation   7
1.5 Schelling grid after one move   8
1.6 Schelling grid after sixteen moves   9
1.7 Segregation after sixteen moves   10
2.1 The union of two sets   15
2.2 Two disjoint sets   16
2.3 One set contains another   16
2.4 The Boolean lattice for the power set of all subsets
    of {x,y,z}   19
2.5 A homomorphic image for the southern women events   22
3.1 The convergence of proportions of heads to probabilities   27
3.2 Probabilities of Type II error   35
4.1 Liking relations among people and "greater than" relations
    among numbers   41
4.2 Respect among the Ricardos and Mertzs   41
4.3 An antisymmetric relation   42
4.4 All possible relations among three people   45
4.5 A relation that is not symmetric, transitive, or reflexive   48
4.6 Structurally equivalent positions in Figure 4.5   48
4.7 An antisymmetric relation   50
4.8 Transitive closure of 4.7   50
5.1 A graph with five nodes and four edges   54
5.2 A network of five persons and four relationships   54
5.3 A complete network   55
5.4 A cycle network   56
5.5 A grid network   57
5.6 A graph that is not simple   57
5.7 A bimodal (bipartite) network   58
5.8 A digraph   59
5.9 A digraph that is a tree   59
6.1 The seven San Francisco Bay bridges   62
6.2 A network with a bridge   62
6.3 A network with local bridges   62
6.4 Networks of strong and weak ties   65

6.5 The forbidden triad   66
8.1 Friendships among a set of workers   77
8.2 A network with two cliques   80
10.1 Friends in the bank wiring room   91
10.2 Marriage network in Renaissance Florence   92
10.3 Example network for betweenness centrality   97
10.4 Network for Exercise 2   100
11.1 Network with two isolates   106
11.2 Clustered circle graph, $n = 10$ 112
11.3 Clustered circle graph with no rewiring, one rewiring, and five rewirings, $n = 50$ 114
12.1 Top: A generic power-law (Pareto) distribution ($N = 10,000$). Bottom: The same distribution, showing only values between 1 and 5   119
12.2 Randomly generated numbers from a power-law (left) and a log-normal (right) distribution   120
12.3 Scale-free network generated by preferential attachment ($N = 1,000$)   124
12.4 Scale-free network generated by the copy model ($N = 1,000$)   125
12.5 Scale-free network generated by the configuration model ($N = 1,000$)   126
12.6 Line and star network ($N = 10$)   132
12.7 Exercise network   136
13.1 Balanced and unbalanced cycles of length 3   138
13.2 Balanced and unbalanced cycles of length 4   139
13.3 Balanced and unbalanced cycles of length 5   140
13.4 The initial conflict situation   141
13.5 Option 1   142
13.6 Option 2   143
13.7 Option 3   143
13.8 Balance exercise   148
14.1 Removing balls from a bowl   150
14.2 A graph representation for party switching   151
14.3 Switching marbles in two bowls   152
14.4 Graph of rat maze   153
14.5 Graph for the penny game   154
14.6 Converging probabilities for Example 2   157
15.1 Life and death as a process   162
15.2 Life cycle of a cat   164
15.3 World countries shaded by life expectancy   167
15.4 Life cycle of a cat (with births)   172
15.5 Population distribution of our cats at equilibrium   178
16.1 Firm A's profit as a function of both firms' expenditures   182
16.2 When TFT does better than unconditional defection   188
17.1 A romantic exchange network with three nodes   191

17.2 A romantic exchange network with five nodes   191
17.3 An exchange network with seven nodes   192
17.4 An exchange network with four nodes   193
18.1 Five realizations of a random walk   204
18.2 Epidemics with different beginning percentages but same
       parameters   205
18.3 Zipf's Law of word frequency   208
18.4 Four distributions, with the same mean   209

TABLES

2.1 All possible patterns of grades   13

2.2 Classification of articles on popular music by year and content   17

2.3 Lyrics of popular music types   17

2.4 The southern women data, 18 women and 14 groups   21

3.1 Payoffs for roulette bets   28

3.2 Movie experience by sex   30

3.3 All sequences of four successes and two failures in six trials and their probabilities   33

3.4 Probabilities of successes of an effective versus ineffective drug   34

4.1 Typical sociology courses   39

4.2 All possible ordered pairs with A, B, and C   39

4.3 Chapter glossary   51

5.1 Graph with 50 nodes   60

5.2 Graph with 100 nodes   60

10.1 Distances between vertices in the bank wiring room   95

10.2 Centrality scores for the bank wiring room network in Figure 10.1   101

10.3 Centrality scores for the marriage network in Figure 10.2   101

11.1 How density decreases in large networks   110

12.1 Attachment probabilities under different systems for Freddy   123

12.2 Distribution of sex partners from the 2008 General Social Survey   134

13.1 Sixty-four triplets classified into 16 MAN triads   147

16.1 The prisoner's dilemma   181

16.2 The roommate's dilemma, outcomes for Roommate A   181

16.3 The prisoner's dilemma, abstractly   181

16.4 A game with two pure and one mixed Nash equilibrium pairs   183

16.5 Strategies in the iterated PD game   187

16.6 Expected earnings for a row player in TFT versus unconditional defection   187

17.1 Coalition excesses under an equal division by three investors   197

17.2 Coalition excesses in the kernel   197

17.3 The kernel for Figure 17.4 if all relations are worth $3,000   198

17.4 The kernel for Figure 17.4 if all relations are worth $3,000, $1,000, and $2,000   199

PREFACE

This book is designed for the undergraduate mostly sociology students we taught at UCLA in courses on mathematical sociology and social networks. They were interested in sociological issues, and they had no background in mathematics beyond high school algebra and the statistics course they were required to take as part of the sociology major. Increasingly, they were aware of the importance of mathematical models and computer simulations in the social sciences, and, of course, over the years they were increasingly comfortable and at ease with the computer. Teaching them a usable amount of calculus would have been a major undertaking, but we knew that interesting uses of finite mathematics were well within their reach.

This book introduces all the mathematics it uses: set theory, the probability function, matrix multiplication, graphs, elementary game theory, groups and their graphs, and Markov chains. Its most distinctive aspect, and what sets it off from earlier books on mathematical sociology and most more recent books on mathematics in the social sciences, is its emphasis on social networks, an area that is clearly the most exciting and fastest growing of mathematical sociology. Moreover, students know they live in an increasingly complex, interdependent, and networked world, and they want to know more about it.

This book is also distinctive in its use of embedded computer demonstrations that are used in the homework and can be used in class. It was the availability of Mathematica Player, a free download available from Wolfram Research, that energized us to rewrite this book. Most of the demonstrations were written by us; a few we downloaded from the Mathematica web site. Player demonstrations do not require that students program. They need only move sliders and press buttons to explore models by varying their parameters. The demonstrations are available from the web and can be run either within Mathematica or with the Mathematica Player, available for free on the web. The simulations are used to increase student understanding of the material, for homework assignments, and, on occasion, to describe models that are intractable mathematically.

All the simulations are available for download at the following website: http://press.princeton.edu/titles/9741.html.

The book teaches all the mathematics that is required. We decided not to include any calculus; the student wanting to use calculus is better served by taking a standard calculus class. We decided not to teach any computer programming. It's difficult to teach both programming and substance in the same course. We decided not to focus on agent-based modeling of complex systems. Our feeling is that one can go pretty far with mathematics alone.

We have found that two different quarter-long courses can be taught from this text, one focusing on social networks and the other not. For the social-network-oriented course we use chapters 1, 4–7, 9–13, and the Afterword. For the mathematical sociology course without an emphasis on networks, we assign chapters 1, 2, 3, 7, 8, 14–16, 18, and the Afterword.

Introduction to Mathematical Sociology

CHAPTER 1

Introduction

Mathematical sociology is not an oxymoron. There is a useful role for mathematics in the study of society and groups. In fact, that role is growing as social scientists and others develop better and better tools for the study of complex systems. A number of trends are converging to make the application of mathematics to society increasingly productive.

First, more and more human systems are complex, in a sense to be described soon. World economies are more and more interconnected. Our transportation and communication systems are increasingly worldwide. Social networks are less local and more global, making them more complex, producing new emergent communication patterns, a positive effect, but which also has made us increasingly vulnerable to pandemics, a negative effect. The Internet has connected us in ways that no one understands completely. Power grids are less and less local and are subject to more widespread failures than ever before. New species are increasingly introduced into local complex ecologies with unexpected effects. Our recent climate change has produced a situation in which it is more and more important to predict the future and the effects of human interventions in the complex system of the global weather. The mapping of the human genome makes available to biologists the possibility of studying the complex system of interactions between genes and proteins. All of these tendencies mean that scientists in a wide variety of areas—computer science, economics, ecology, genetics, climatology, epidemiology, and others—have developed mathematical tools to study complex systems, and these tools are available to us sociologists.

The second important trend is the growing power and ubiquity of the computer. Computer simulations and mathematics are complementary tools for the study of complex systems. They are two different ways of drawing implications for the future from what is known or assumed to be true. Mathematics can be used to draw far-reaching and sometimes unexpected conclusions using logic and mathematics. For example, many properties of networks have been proved to be true by mathematicians using traditional mathematical tools. Computer simulations use computer programs the coding of which embodies assumptions and whose conclusions are evident after the program has iterated. Computer simulations are useful in situations that are unsolved or intractable mathematically.

This text uses both mathematics and computer simulations. Sometimes the computer simulations demonstrate phenomena for which there is no exact mathematical solution. More frequently simulations are used to illustrate some model so that you, the reader of this book, will gain some understanding of how the model works and how it is affected by varying parameters even if a full mathematical treatment of the model is beyond the purpose of this book.

EPIDEMICS

At the time we are writing this chapter there is an epidemic of concern over swine fever, a variant of influenza that seems to have captured the public's attention. Both the flu and fear of this flu spread through social networks, and we want now to illustrate some of the properties of epidemics through a very simple model. The model will be illustrated both with a little simple mathematics and with a computer simulation.

Suppose that a large population consists of N individuals. Suppose that each individual in the population has small probability p of being connected to each of the others in the population and that his connection to one individual has no bearing on his connection to any other person in the population. This creates a *random network* among the members of the population. Real social networks are not like random networks, but random networks are very tractable mathematically, and so they tend to be assumed by epidemiologists who study the spread of diseases. The powerful conclusions may be relatively unaffected by the unreality of the assumptions, much like the statistician may on occasion assume a normal distribution because the conclusions are not affected very much if the assumption of a normal distribution is not exactly true. We will examine how real social networks differ from random networks in much more detail in later chapters, but for the moment we will assume that they are useful descriptions of real networks.

Despite its unrealism, let's, for the sake of convenience, assume that the network is random. Suppose that initially just one person is sick with a contagious disease. If p is the probability this person is connected to anyone else, then we should expect this person to be connected to $p \times N$ others, on the average. If p is small and N is large, then each of these pN individuals will be connected to pN others, and so the sick individual will be connected to $(pN)^2$ individuals indirectly, through his contacts. If p is small and N is large we can ignore the unlikely event that some individuals will be connected to more than one of his direct connections. Similarly $(pN)^3$ persons will be connected even less directly, through two intermediaries, and $(pN)^{k+1}$ persons will be connected through k intermediaries. Let's look at this sequence:

$$1 + pN + (pN)^2 + (pN)^3 + (pN)^4 + \cdots \tag{1.1}$$

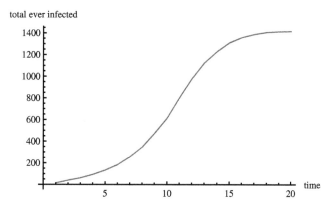

Figure 1.1. Total number of infected individuals when $pN = 1.5$.

This represents the expected or average number of people who get the contagious disease carried by one person. Of course, this number must in fact be limited because the population is of finite size N, but if $pN \geq 1$, this sum diverges: it just gets bigger and bigger, without limit. If $pN < 1$, however, the sum converges to a number, and this number may be quite small relative to N. The sequence in equation 1.1 converges to,

$$1 + pN + (pN)^2 + (pN)^3 + (pN)^4 + \cdots = \frac{1}{1 - pN} \qquad (1.2)$$

You can verify this for yourself by substituting a few values. When $pN = .5$, for example, the sums are 1, 1.5, 1.75, 1.875, and so on, getting closer and closer to 2, the limit. What this means is that if an infected person infects, on the average, less than one other person the disease will not become an epidemic affecting nearly everyone, but otherwise it will spread to the entire population.

The following figures were generated from a simulation based on a few simple assumptions. The network was of size 2,500. Ten individuals were initially infected. The probability that any two people were connected, also the probability that an infected person would infect a healthy person in any given time period, was set at .0006: $p = .0006$. The network was examined over 100 time periods. If infected in one time period, the person was assumed to be infectious at the next time period and immune thereafter. In this case $p \times N = .0006 \times 2,500 = 1.5$, and 1.5 is bigger than the critical value of 1.00 ties per person. We should expect the disease to spread.

Figure 1.1 shows the total number infected. After just 25 time periods most had been infected. The curve shows a familiar S-shaped figure, called the logistic curve. The disease spread slowly when few had it, then picked up speed, then slowed down as there were fewer and fewer who had not been infected and were not immune. Figure 1.2 shows the number of new cases in each time period, telling the same story as Figure 1.1 in a different way.

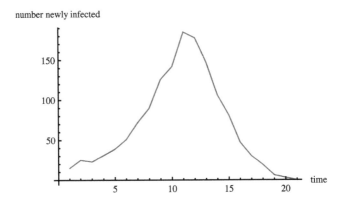

Figure 1.2. Number of newly infected individuals over time when $pN = 1.5$.

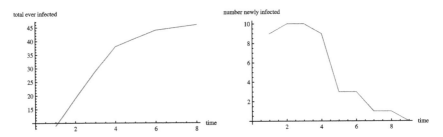

Figure 1.3. Number of total and newly infected individuals over time when $pN = .75$.

Now suppose we make the parameter p half as big, $p = .0003$, so that each individual averages .75 contacts instead of 1.50. This is below the critical value of one contact per person on average. We would expect the outbreak not to become an epidemic and to peter out after the initial set of infected individuals fail to reproduce themselves. Figure 1.3 shows that the total number never passes 50 and illustrates the declining number of new cases.

In later chapters we'll see how these inferences are a consequence of deep and nonobvious properties of random networks. We will also see how the inferences must be modified for other classes of nonrandom networks. Even with these qualifications, the results are interesting. First, they apply to phenomena other than the spread of disease. Information and rumor transition can also be modeled. Coleman et al. (1966) used this model to account for diffusion in the use of a new antibiotic among doctors in Midwestern communities. In this case, what was being spread was not disease but information and influence. The existence of a critical point has policy implications in epidemiology. It means that not everyone need be inoculated for an inoculation campaign to be effective. It also helps explain why Apple computers are less subject to viruses than are PCs. Since computer viruses are targeted for one operating system, any Mac virus will spread only from Mac to Mac, while almost all the computers a Mac is connected to will be PCs. The Mac to Mac network will be very sparse while the PC to PC network will have a lot higher *density* (a higher value of *p*). Thus, the lower frequency of Mac infections need not be due to any superiority of the Mac operating system but simply due to the fact that very few people use Macs.

The spread of a disease or information depends not only on the density of the network but also on the presence or absence of long-range connections in the network, and this topic can be examined both mathematically and through the use of computer simulations. When in history almost all ties between people were strictly local, epidemic diseases spread much more slowly. The "Black Death," a plague that decimated Europe in the Middle Ages, was carried by sailors from Asia to Italian port cities in 1347, but it did not reach England until 1349 and Scandinavia until 1350. This slow spread occurred because long-distance movements hardly existed. Most people never saw anyone outside their own small village. Nowadays worldwide influenza pandemics occur every year.

Simulated Epidemics

Let's explore this difference using the demonstration *Simulated Epidemics*. This demonstration offers the possibility of examining contagion in two different types of networks, random networks and grids. In grid networks all ties are local, like in a farming community where farmers have relations only to those in neighboring farms; there are no long-distance ties. In a random network there are no constraints on the ties at all. In the demonstration, the number of individuals and connections are the same in a grid and random network: there are 400 individuals and the average number of connections is four. Play with the demonstration for a while and you see that there are two major easily observable differences. First, the diseases spread much more rapidly in the random network. Second, the shape of the curves for new cases is quite different. In the grid the number of new cases goes up in a straight line until the edges of the grid are reached. In the random network the number of cases seems to go up exponentially at the beginning.

Why are there these differences? In a grid the disease can expand only at the circumference of the infected area. It is only individuals at the border of the infected area who come in contact with uninfected individuals. On the other hand, in a random network each newly individual in the early stages is coming in contact with four uninfected individuals. In a random network we will have 1 infected individual, then 4, then each of these 4 will infect 4 others so that there will be 16 new cases, then each of these 16 cases will infect 4 more for 64 more cases, and so on. The number of new cases will not be proportional to time, t, but 4^t.

RESIDENTIAL SEGREGATION

Residential segregation by race and class has many causes. In the past some of it was legal: residential covenants prevented whites from selling their homes to nonwhites. Patterns established in this way persist. Some segregation is economic because race and income are correlated and neighborhoods with different priced homes may become segregated primarily by income but also incidentally by race. However, some segregation patterns result from individual choices: individuals may wish to avoid being a minority in their own community, or at least being a small minority. Such voluntary segregation based on the desire to be in proximity to similar others can produce segregation by sex or social class as well as race. In a lunchroom of young children or at a party among middle-aged adults, there may be segregation by gender. People of different social classes don't meet each other in unstructured situations, but our experience in public schools was that school social events and cafeteria lunch groups were segregated by class, sex, and race.

Thomas Schelling (1969, 1972), an economist, was the first to explore this phenomenon in simulations. Schelling devised what would now be called an agent-based model. In this simulation individual actors were placed in a two-dimensional grid, an eight by eight checkerboard. Actors were of two different types: "X" and "O" actors. Each actor could change his position, following certain simple rules. Each actor wanted a minimum percentage of her neighbors to be of the same type. On this grid each actor had up to eight neighbors: to the right, left, above, below, upper right, upper left, lower right, and lower left. Actors at the corners of the grid had only three neighbors, while actors on an edge but not a corner had five neighbors. Some squares of the grid would be empty, to permit movement.

Each actor followed a simple rule: he required that a certain minimum proportion of his neighbors be of his type. Suppose, for example, that desired minimum proportion were greater than one-third. Then if a person had just one neighbor, that actor would have to be of the same type; if two neighbors, at least one must be of the same type; if three, four, or five neighbors, at least two; and if six, seven, or eight neighbors, at least three. If an actor were dissatisfied she would move to the nearest position on the grid that was satisfactory in terms of the composition of the neighborhood.

The interesting thing about such situations is that they can lead to cascades of movement. If one actor moves he shifts the character of his previous neighborhood, increasing the prominence of the other type, so that others who were satisfied before his move may, as a consequence of his move, become dissatisfied with the composition of their neighborhood. These cascades are *emergent properties* of the simulation: there is usually no simple way of predicting the outcome. Schelling showed that even such simple rules could have unexpected outcomes.

Schelling presented this model in 1972, when computers were not as prominent in research as they are today. He used a small eight-by-eight grid

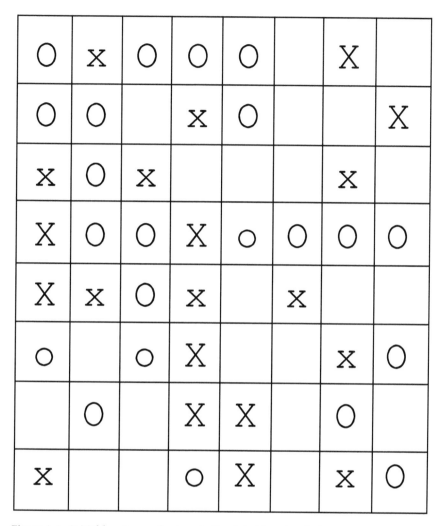

Figure 1.4. Initial locations of actors in the Schelling simulation.

that he could manipulate by eye, instead of the much larger grid he could have used even with the computers of his day. However, there is another effect of his choice not to use computers. Using a checkerboard-sized grid and moving the pieces by eye emphasized his larger point—that simple assumptions, easily implemented, can produce unexpected outcomes.

Of course, there are outcomes that are not surprising. If neither type is willing to be a minority in its neighborhood, the result will be complete segregation of the pieces, each type occupying its own separate areas of the grid. But suppose that each actor wants her own type to constitute more than a third of her neighbors. Figure 1.4 shows an initial distribution with about one-third X type, one third O type, and one third empty squares.

O	x	O	O	O		X	
O	O	X	x	O			X
	O	X				x	
X	O	O	X	o	O	O	O
X	x	O	x		x		
O		o	X			x	O
	O		X	X		O	
X			o	X		x	O

Figure 1.5. Schelling grid after one move.

Squares in which one-third or less of the neighbors are of the same type (the dissatisfied actors) are shown in a smaller font.

In all, 27 actors are satisfied with their neighborhoods and 15 are not. On the average, 45% of anyone's neighbors are of the same color. Suppose that one of the unhappy actors moves. For example, the X actor in the third row and first column has five neighbors, only one of whom is a fellow X. A move to the square in the second row and third column will give him three X and three O neighbors, 50% with his type. Figure 1.5 shows the grid after his move.

As a result of this move, not only is he (the actor initially on (3,1) who moved to (2,3)) happier, but the actor in the third row and third column now also has more than one-third of his neighbors of his type.

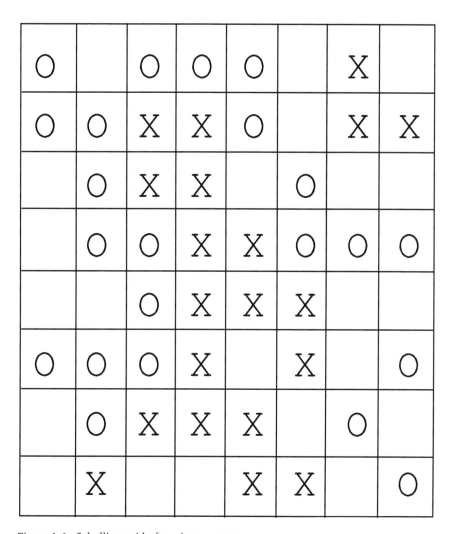

Figure 1.6. Schelling grid after sixteen moves.

Figure 1.6 shows the results after 16 moves. Note that all actors live in acceptable neighborhoods: no letters are in small fonts. The degree of segregation into primarily X and O regions is quite evident. One possible measure of segregation is the average proportion of neighbors who are of the same type. Figure 1.7 shows that this measure climbed regularly. By the sixteenth move the degree of segregation is almost three-quarters, even though the actors are programmed to require that more than one third of their neighbors be of their own color. This is one of the surprising aspects of this simulation—the high degree to which the actual segregation exceeds the minimum desired degree. This high proportion is clearly an emergent property of the simulation. As Schelling points out, no one may want to live in such a highly segregated society. Everyone may actually prefer more

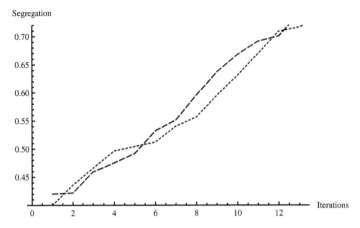

Figure 1.7. Segregation after sixteen moves.

integration, but a highly segregated society can emerge from independent actions by unwitting actors who do take the larger context into account.

Schelling's Model of Residential Segregation

This demonstration enables you to examine Schelling's model and to vary its parameters. You can change the relative sizes of the majority and minority (represented by blue and green squares), the proportion of empty squares, and the preferences of the majority and minority for having neighbors of their own color. Some homework problems also use this demonstration. In this chapter we have tried to give you, the reader and teacher, a sense of what this book is like. No mathematical background beyond simple algebra is assumed. All additional mathematics will be introduced as needed. The tools are simple mathematical models and computer simulations of social processes. Computer simulations are used when the phenomenon itself is intractable mathematically and to give you the reader a sense of how a mathematical model develops and is sensitive to changes in its parameters. The emphasis is on conclusions that would not be obvious without the use of mathematics or simulations. These are the unexpected emergent properties of social systems of complexly interconnected individuals, and it is the study of these emergent properties that constitutes the bailiwick of sociology.

Chapter Demonstrations

- *Schelling's Model of Residential Segregation* simulates the Schelling model of residential segregation.

- *Simulated Epidemics* simulates the spread of an epidemic in a random grid network. The density of contagious connections in the network can be varied.

EXERCISES

1. The simulation *Simulated Epidemics* allows for variation in the size of a population living on a square grid (the size of the population is the squared length of a side of the grid) and the probability of a random connection between any two nodes, labeled "Density of Network." Below the critical value of the density, the disease dies out and above that it becomes an epidemic. What are the corresponding critical values of the density for each grid? Determine these results theoretically from what the chapter says about critical values.

2. Verify these answers by showing, with *Simulated Epidemics*, that values 50% larger than the critical value of the density produce epidemics that spread to most of the population.

3. For a 50 by 50 grid, what would you expect to be the effect of network density on the time it takes for the epidemic to reach 80% of the population? Would increasing the density increase or decrease the time it takes for the epidemic to infect 80% of the population? Provide evidence for your conjecture by running the simulation with different densities.

4. Using the *Schelling* demonstration, create a community in which 50% of the squares are green, 25% are blue, and both types prefer that at least one-third of their neighbors are the same type. Move unhappy squares as long as you can (the simulation will stop when no square is dissatisfied with its neighborhood). Do this a few times. The segregation index is simply the average proportion of same-color neighbors. Although people are willing to live in neighborhoods with one-third of the other color, what is the average value of the index?

5. Now change the last simulation so that both types of squares want to live in neighborhoods in which their color is in the majority. What happens to the segregation index?

CHAPTER 2
Set Theory and Mathematical Truth

Individuals' memberships in groups are an important kind of sociological data. For example, there has been much analysis of interlocking directorates among major corporations. If two firms share a director, there is a pathway for communication and coordination between the firms. Clusters of firms that might share interests and coordinate their behaviors might also be revealed by the presence of interlocking directorates (Roy and Bonacich, 1988). Especially central firms that overlap with a variety of other firms and are in a position to coordinate activities in the entire community might also be revealed. Bonacich and Domhoff (1981) examined patterns of overlap among elite social clubs and policy groups in the United States to see if there were clusters and if there were clubs and policy groups that transcended local geographical boundaries (there were). When sociologists look at the relations between memberships in groups, they are implicitly using set theory, which we will introduce in this chapter.

In sociology, and in everyday life, we often categorize things in a variety of ways. Mathematicians have a special name for such groupings. They call them sets. Mathematicians have developed special symbols to describe sets and things one does with sets.

Definition. *Sets* are collections of things.

A set can be described in two ways—either by listing the elements in the set or by describing the property that defines the set. For example, the set S could be the set of whole numbers between 1 and 10. We could express this in two ways:

Example 1

$S = \{1, 2, 3, 4, 5, 6, 7, 8, 9, 10\}$

$S = \{x | x$ is a whole number greater than or equal to 1 and less than or equal to 10} [This is read as "The set of numbers x such that x is a whole number greater than or equal to 1 and less than or equal to 10."]

These two definitions define the same set S.

Table 2.1.

All possible patterns of grades

Element in S	Examination	Presentation
s_1	A	P
s_2	B	P
s_3	C	P
s_4	D	P
s_5	F	P
s_6	A	NP
s_7	B	NP
s_8	C	NP
s_9	D	NP
s_{10}	F	NP

Example 2

Or, consider another possibility. Suppose that there is one examination in a course and one class presentation graded P or NP. Let the set S consist of all the possible different patterns of grades. The set has 10 elements, S_1 through S_{10}. These 10 elements and the combinations they represent are given in Table 2.1.

There is a special symbol for an element of a set: "$\in$". If a is an element of the set S we write $a \in S$. In Example 1 above, $3 \in S$. In the Example 2, $S_{10} = (F, NP) \in S$.

One set T can contain another set S in the sense that all members of S are also in T. All students in sociology course (set S) are enrolled at their college (set T).

Definition. Set S is a *subset* of set T, written $S \subseteq T$, if every element in S is also in T (or, if $x \in S$, then $x \in T$).

For example, let S be the set of whole numbers between 1 and 10 inclusive (including the endpoints 1 and 10) and let T be the set of all whole numbers. Then $S \subseteq T$. In this case we can say something even stronger. All the elements of S are contained in T and there are elements of T that are not in S. When this is true, S is said to be a proper subset of T, and we write $S \subset T$. Thus, the bar underneath the symbol "$\subseteq$" in the symbol "$\subset$" provides for the case where S may be identical to T.

Suppose that a student passes the course if she gets a C or better in the examination or receives a D in the examination and a P in the presentation. These outcomes are a proper subset of S consisting of the patterns {(A,P), (B,P), (C,P), (D,P), (A,NP), (B,NP), (C,NP)}.

Sometimes we want to refer to all those who are members of two different sets. We might want to study, for example, female minority students or business people who are on the boards of directors of both General Motors and IBM.

Definition. The *intersection* of two sets S and T, written $S \cap T$, consists of all elements in both sets: $S \cap T = \{x | x \in S \text{ and } x \in T\}$.

For example, let A be the set of all whole numbers greater than or equal to 10 and let B be the set of all whole numbers less than or equal to 19. Then $A \cap B$ is all the whole numbers from 10 through 19. Similarly, if A is the set of all currently registered college undergraduates and B is the set of all left-handed humans, then $A \cap B$ is the set of all left-handed currently enrolled college undergraduates.

Suppose that A is the set of all even numbers and B is the set of all odd numbers. It's clear that no numbers belong to both sets. Rather than say that $A \cap B$ does not exist, we say that $A \cap B$ is empty. There is, moreover, a special symbol, Ø, for the empty set. For the sets of even and odd numbers we write $A \cap B = $ Ø. When $A \cap B = $ Ø we say that A and B are disjoint or mutually exclusive sets.

Sometimes we want to refer to all those who are members of one set or another. We may want to know how many faculty members in a university are female or minority.

Definition. The *union* of two sets S and T, written $S \cup T$, consists of all elements that are in S or T (or both): $S \cup T = \{x | x \in S \text{ or } x \in T\}$.

For example, let S be the set of all even numbers, T the set of all odd numbers, and J all the whole numbers. Then $J = S \cup T$. If A is the set of all women teaching at a university and B is the set of all minority faculty, then $A \cup B$ is the set of all faculty who are women or are members of minorities (or both).

Let S be a set consisting of three elements: $S = \{a, b, c\}$. We had previously defined a subset of S as a set contained entirely in S. A little thought will show that there are eight subsets of $S = \{a, b, c\}$. These are

- $S_1 = \{a, b, c\}$
- $S_2 = \{a, b\}$
- $S_3 = \{a, c\}$
- $S_4 = \{b, c\}$
- $S_5 = \{a\}$
- $S_6 = \{b\}$
- $S_7 = \{c\}$
- $S_8 = $ Ø (The empty set is contained in every other set, Ø $\subseteq S_i$ for every i.)

In general, if a set has n elements, there are 2^n subsets. Thus, if a set consisted of 4 elements, there would be 16 subsets (or 2^4 subsets). The set of all the subsets of the set S is called the *power set* of S. It is written as 2^S.

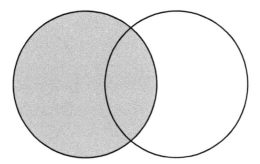

Figure 2.1. The union of two sets.

The power set is a set of sets. With a three element set S, 2^S is the following set:

$$2^S = \{S_1, S_2, S_3, S_4, S_5, S_6, S_7, S_8\} \tag{2.1}$$

Definition. Finally, let A be a set. A', the *complement* of A, consists of all elements not in A: $A' = \{x | x$ is not in $A\}$.

If A is the even numbers, then A' is the odd numbers. It's not always clear what the complement of a set is. For example, if S is the set of all UCLA undergraduates, what is S'? The set of UCLA graduate students? USC undergraduates? All non-UCLA undergraduate students in the United States? All LA residents who are not UCLA undergraduates? The choice of complement sometimes makes a difference in research. For example, if one were doing a study of political attitudes, it would make a difference which of the above groups one compared UCLA student attitudes to. UCLA might seem liberal relative to USC students but illiberal in comparison to UCLA graduate students.

The complement of a set cannot be unambiguously defined without reference to a *universal set U*. The complement of a set S consists of all the members of the universal set U who are not in S. If S is the set of all UCLA undergraduates and U is the set of all UCLA students, then S' is the set of UCLA graduate students. But, if U is the set of all Los Angeles college undergraduates, then S' contains USC undergraduates, CSULA undergraduates, and so on.

Relations between sets can be pictured using Venn diagrams. In a Venn diagram sets are represented by the areas within circles. The two circles in 2.1 represent A and B. The dark doubly lined area in the middle is $A \cap B$. The area enclosed by the two circles is $A \cup B$.

Figure 2.2 represents the situation in which sets A and B are disjoint: $A \cap B = \emptyset$.

Figure 2.3 shows the Venn diagram for when set B is a proper subset of set A.

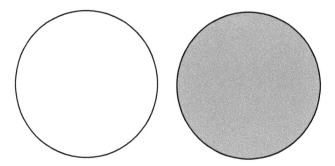

Figure 2.2. Two disjoint sets.

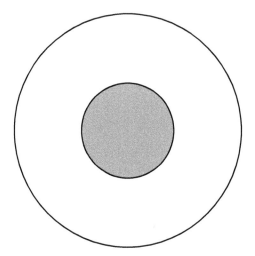

Figure 2.3. One set contains another.

The Mathematica demonstration *Venn Diagrams* and *Set Intersection and Union* enable you to explore further the properties of Venn diagrams. Note that the diagram of the three sets in *Venn Diagrams* has eight areas and that seven of them are either colored or not. How can the eighth area, outside the three circles, be described? *Set Intersection and Union* allows you to vary the sizes of the three circles whose inner areas represent sets. Can you draw a diagram in which one of the circles contains the other two? Why is the two-way intersection just the two smaller circles?

Relations between sets can also be represented by tables, or cross-classifications, as they are known in statistics. Table 2.2 is a classification of articles in various mainstream magazines and newspapers (the *New York Times, Time, Newsweek*, etc.) discussing heavy metal or rap music in the years 1985 through 1990 (Binder, 1993).

Table 2.2 is a classification of 108 articles. The rows (years) represent mutually exclusive subsets of the articles because no article was published

Table 2.2.
Classification of articles on popular music by year and content

Year	Discuss heavy metal but not rap (H)	Discuss rap not heavy metal (R)	Discuss heavy metal and rap ($H \cap R$)
1985	13	1	1
1986	1	5	0
1987	4	7	0
1988	6	7	0
1989	5	5	5
1990	5	33	10

Table 2.3.
Lyrics of popular music types

Content of lyrics	Heavy metal	Rap
Hard swear words	2	9
Sex, graphic	1	7
Violence or murder of police	0	2
Rebellion against teachers/parents	2	0
Degradation and violence to women	3	6
Sex, indirect references	2	1
Grisly murder, violence, torture	1	0
Sex, group	2	1
Drugs and/or alcohol	1	0
Incest	1	0
Prejudicial slurs	1	1
Suicide	1	0

in more than one year. Similarly, the columns represent another set of mutually exclusive subsets of the articles. We can see from the table that there are $13 + 1 + 4 + 6 + 5 + 5 = 34$ articles in the set H (heavy metal). The cell values show the number of articles in the intersections of various sets. For example, there are 33 articles in the set $1990 \cap R$.

Unions of subsets are combinations of rows, combinations of columns, or combinations of both rows and columns. There are 21 articles in the set $1985 \cup 1986$ (a combinations of columns), 92 articles in the set $H \cup R$ (a combination of two columns), and 77 articles in the set $1990 \cup H$ (a row and a column).

The same researcher attempted to classify the lyrics of 10 often controversial heavy metal and rap songs in the same period (1985–1990). The results are given in the Table 2.3.

This table is quite different from the previous one. It is not a cross-classification. The rows are not disjoint categories. A song could fall into two or more of the row categories. For example, it could use hard swear words and talk about drugs. This explains why the sums of the two columns are not 10 each despite the fact that only 10 heavy metal and 10 rap songs were analyzed.

There are rules for using these three symbols, "∩", "∪", and "'". Using Venn diagrams you should be able to convince yourself that the following rules hold for these operations.

1. Idempotency of ∩ and ∪

   (a) $A \cap A = A$, $A \cup A = A$

2. Commutivity of union and intersection operations

   (a) $A \cap B = B \cap A$
   (b) $A \cup B = B \cup A$ [Note: What this means is that you get the same set $A \cap B$ if you take the members of set A that are also in set B or if you take the members of set B that are also in set A. The same holds for unions.]

3. Associativity of union and intersection operations

   (a) $(A \cap B) \cap C = A \cap (B \cap C)$
   (b) $(A \cup B) \cup C = A \cup (B \cup C)$ [Note: What this means is that you get the same set $(A \cap B) \cap C$ if you take the members of sets A and B $(A \cap B)$ that are also in C or if you take the members of B and C, $B \cap C$, that are also in A. The same holds for unions.]

4. Distributivity

   (a) $A \cap (B \cup C) = (A \cap B) \cup (A \cap C)$
   (b) $A \cup (B \cap C) = (A \cup B) \cap (A \cup C)$ [Note: Convince yourself of the truth of these two statements using a Venn diagram for three sets.]

These properties are familiar because they also hold true for ordinary arithmetic. Substituting " + " for "∪" and "×" for "∩," addition and multiplication are both commutative and associative, and the first (but not the second) type of distributivity is also true. The second kind of distributivity is a special property of set union and intersection.

Together with a few other properties, these define what is called a *Boolean algebra*, named after the 19th-century British mathematician George Boole. These principles can be used to simplify complicated expressions and to show that various sets are equal to each other. For example, you can show that $(A \cap B)' = A' \cup B'$ and $(A \cup B)' = A' \cap B'$ for any sets A and B, which are useful and nonobvious facts.

The elements of a Boolean algebra can be drawn in the form of a *Boolean lattice*, a hierarchical diagram in which set A is below set B if $A \subset B$, $A \cap B$ is the unique highest element below both A and B, and $A \cup B$ is the unique

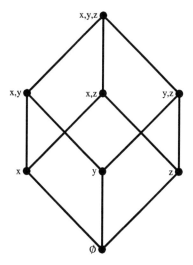

Figure 2.4. The Boolean lattice for the power set of all subsets of {x,y,z}.

lowest element above both *A* and *B*. For example, consider the power set for three objects x, y, and z. Figure 2.4 shows the Boolean lattice for the power set of all subsets of {x,y,z}.

The demonstration *Boolean Algebra* shows the Boolean lattices for the power sets of sets of varying sizes.

BOOLEAN ALGEBRA AND OVERLAPPING GROUPS

One form of data in sociology concerns membership in groups. Sociologists have made careers from studying patterns of interlocking directorates among the largest corporations. Sharing directors is a type of potential communication and coordination link between corporations, and serving on boards together is a potential link between directors. The resulting networks of directors and corporations can be explored for patterns. Which corporations cluster together? Are there particularly important corporations or directors that are particularly central and important and tie the economy together? In high schools are there clubs that bring together especially active students? Are there students who are active in the most central clubs? Analysis of the Boolean structure might be one way to go. Groups with similar memberships, located near each other in the Boolean lattice of the power set, are similar to each other in membership and might form a cohesive cluster of firms with similar boards of directors. Companies that are placed high in the Boolean lattice will contain, on the other hand, a variety of members and might be thought of as being particularly central. Thus, position in the Boolean lattice might give us a clue about clusters and centrality.

But, to analyze data of this sort we must simplify them in some way. With just 20 people there are more than one million elements in the power set, and even a drawing of the Boolean lattice becomes impractical. A desirable simplification would have the following properties, if we are to continue to use the Boolean lattice as a visual description of the relationship between groups. We want a smaller Boolean lattice that preserves relationships between groups: if, for example, two groups are disjoint, they should be disjoint in the smaller image Boolean lattice. If one group is contained in another, it should be contained in the image lattice as well.

To do this we'll make use of the mathematical concept of a *mapping*. This may first seem like a round about way of expressing simple ideas, but there are some advantages. A mapping is a relationship between two sets in which elements of the two sets are matched with each other. A *many-to-one* mapping is one in which many elements of a larger set are matched with each element of the smaller set. For example, we might match a large set of high school students with a smaller set of terms representing ethnic groups. Our hypothesis might be that ethnic relations are very important in determining interpersonal relations in the high school. If the mapping is a good one, then relationships between students should be preserved by the mapping: if two students like one another then they should be mapped into the same ethnic group; if two students dislike one another, they should be mapped into different ethnic groups. To the extent that this is true we can characterize the interpersonal relations between students as a set of relations among ethnic groups.

Let L be a Boolean lattice and I the smaller, easier to interpret image lattice. Let h be a mapping from one to the other showing how groups in the larger lattice are represented in the smaller and more comprehensible lattice. If G is a set in L and a member of lattice L, then $h(G)$ is a member of I. Our mapping h will have the following properties:

1. $h(A \cap B) = h(A) \cap h(B)$
2. $h(A \cup B) = h(A) \cup h(B)$

Thus, the largest group that is a member of both groups A and B, $A \cap B$, is mapped into the highest element in I that is below both $h(A)$ and $h(B)$, and similarly for unions. Such a mapping can be shown to preserve the order of the elements also: if $A \subseteq B$, then $h(A) \subseteq h(B)$ as well; if A and B are disjoint, then $h(A)$ and $h(B)$ will be disjoint also. Such a mapping is called a homomorphism, and what we are looking for is a useful homomorphism.

The following table shows the attendance of 18 women at 14 events in a small southern town. The rows and columns have been permuted to reveal a clear pattern. There are two groups of women: 1 through 9 and 10 through 18. There are three kinds of events: 1–6, attended by the first set of women; 10–14 attended primarily by the second set; and 7 though 9

Table 2.4.
The southern women data, 18 women and 14 groups

	1	2	3	4	5	6	7	8	9	10	11	12	13	14
1	1	1	1	1	1	1	0	1	1	0	0	0	0	0
2	1	1	1	0	1	1	1	1	0	0	0	0	0	0
3	0	1	1	1	1	1	1	1	1	0	0	0	0	0
4	1	0	1	1	1	1	1	1	0	0	0	0	0	0
5	0	0	1	1	1	0	1	0	0	0	0	0	0	0
6	0	0	1	0	1	1	0	1	0	0	0	0	0	0
7	0	0	0	0	1	1	1	1	0	0	0	0	0	0
8	0	0	0	0	0	1	0	1	1	0	0	0	0	0
9	0	0	0	0	1	0	1	1	1	0	0	0	0	0
10	0	0	0	0	0	0	1	1	1	0	0	0	0	0
11	0	0	0	0	0	0	0	1	1	1	0	1	0	0
12	0	0	0	0	0	0	0	1	1	1	0	1	0	0
13	0	0	0	0	0	0	1	1	1	1	0	1	1	1
14	0	0	0	0	0	1	1	0	1	1	1	1	1	1
15	0	0	0	0	0	0	1	1	0	1	1	1	0	0
16	0	0	0	0	0	0	0	1	1	0	0	0	0	0
17	0	0	0	0	0	0	0	0	1	0	1	0	0	0
18	0	0	0	0	0	0	0	0	1	0	1	0	0	0

attended by members of both sets. What type of Boolean lattice image can be created to model this pattern?

It is easy to show (famous and misleading words in mathematics) that homomorphisms of Boolean lattices involve selecting a small subset of individuals and clustering subsets from the larger set according to the membership patterns of this smaller subset. Let us select two individuals, Women 3 and 13, to create the homomorphic image. In 2.5 Woman 3 attended the events on the left, Woman 13 attended the events on the right, and both women attended Events 7, 8, and 9. Neither woman attended Events 1 or 11. We might conclude that Events 7, 8, and 9 were the most integrative and cohesion building because they were attended by women from both cliques.

TRUTH AND FALSITY IN MATHEMATICS

For most of us the truth is simply that which is true most of the time. When we say that "Catholics believe in the infallibility of the Pope" we don't mean that every single Catholic believes this, just that most do. When we say that those with a college education make more money than those without, we don't mean that this is true without exception. Mathematicians have an unfamiliar conception of what a true statement is. For mathematicians a true statement in mathematics is one for which

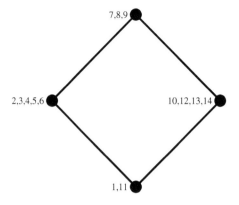

Figure 2.5. A homomorphic image for the southern women events.

there is *no* exception. As we shall see, this conception of the truth is both stronger and weaker than the usual conception of what is true.

Take the following examples. The reciprocal or inverse of a number is 1 divided by that number. The reciprocal of 5 is 1/5. Is it true that "every number has an inverse"? No, because zero does not have an inverse. Almost all numbers have an inverse, but because zero does not, the statement is false. On the other hand, the statement that "every whole number is even or odd" is true because there is not a single exception. There are some statements whose truth is as yet undetermined. Take for example the apparently true statement that every even number is the sum of two prime numbers. Mathematicians using computers have not yet found an exception to this rule, but they are not sure that if they continued to look they might find one someday.

Mathematicians must have such a strict definition of "truth" because they prove things through the use of logic, and logic works best when you are dealing with statements that are always true. Mathematics also contains a rigorous definition of a proof. Many mathematical statements are of the form "A implies B" or "if A then B," where A is a set of assumptions and B is a conclusion. We will first look at a nonmathematical example of this logic. Suppose we want to "prove" that all sociologists subscribe to the *American Sociological Review*. We could also express this idea in the "if ... then ..." format: if someone is a sociologist, then she *subscribes* to the *American Sociological Review*.

Now we could prove this statement in two different ways. We could survey sociologists and see if all of them subscribe, or, equivalently, we could survey all academics who do not subscribe to the *ASR* and see if we find a sociologist. Either survey will test the statement, because both surveys will locate the only people who can disprove the statement: sociologists who do not subscribe to the *American Sociological Review*. So, the statement that "if A then B" ("if an academic is a sociologist, then he subscribes to the *ASR*") is equivalent to the statement "if not B, then not A" ("if someone does not subscribe to the *ASR*, then she is not a sociologist"). "If A then B" is true under exactly the same circumstances as "if not B, then not A."

We can also put this statement in the language of set theory. Let S be the set of academic sociologists and let A be the set of subscribers to the *ASR*. The assertion that all academic sociologists subscribe to the *ASR* is equivalent to the claim that $S \subseteq A$ or, equivalently, $A' \subseteq S'$.

This mathematical criterion of truth—that something is true if there are no exceptions—can have some paradoxical implications for the nonmathematician. For example, is it true or false that all unicorns are fond of apples? This statement is false only if there are exceptions—unicorns that do not like apples. There are no exceptions. Therefore, it is true: unicorns like apples. Similarly, it is also true that unicorns hate apples. As we shall see in the next chapter, there are reasons for sticking with the mathematician's definition of truth in this course.

Chapter Demonstrations

- *Venn Diagrams* labels various subsets of three sets using Venn diagrams
- *Set Intersection and Union* looks at the relations between two sets
- *Boolean Algebra* looks at homomorphisms of Boolean lattices

EXERCISES

1. [From Kemeny et al. (1966)] Joe, Jim, Pete, Mary, and Peg are to be photographed. They want to line up so that the boys and girls alternate. List the set of all possibilities.
2. [From Kemeny et al. (1966)] In Problem 1, list the following subsets.

   (a) The set in which Pete and Mary are next to each other.
   (b) The set in which Peg is between Joe and Jim.
   (c) The set in which Jim is in the middle.
   (d) The set in which Mary is the middle.
   (e) The set in which a boy is at each end.

3. Let $S = \{$Andy, Charles, George, Pete$\}$, $T = \{$Pete, Harry, Barbara, Cynthia, Harriet$\}$, $V = \{$Philip, Pete, Anne, Andy$\}$, and the universal set $U = S \cup T \cup V$. Who belongs to the following sets?

   (a) $S \cap T$
   (b) $S \cap T \cap V$
   (c) $(S \cup T)'$
   (d) $V \cap T'$
   (e) $V \cup T'$

4. If $A \cup B = \emptyset$, what can you conclude about A and B?

5. If $A \cap B = A$, what can you conclude about the relation between A and B?

6. If $S \cup T = S$, what can you conclude about the relation between S and T?

7. Is it always true that $A \cap B \subseteq A \cup B$ for any sets A and B?

8. What is true about the set $A \cap A'$?

9. Is it true that $(A')' = A$ (in words, the complement of the complement of a set is the set itself)?

10. Let A and B be any sets. Is it true or false that $(A \cap B)'$ is not equal to $A' \cap B'$. Is it true or false that $(A \cap B)' = A' \cup B'$. [Note: The left side of this equation refers to the complement of the set $A \cap B$.]

11. Is it true or false that if $A \subseteq B$, then $B' \subseteq A'$.

12. The equations $y = 5x + 7$ and $y = 5x + 10$ define lines in a graph. Show that these two lines never intersect.

13. [From Kemeny et al. (1966)]. In testing blood, three types of antigens are looked for: A, B, and Rh. Every person is classified doubly. He is Rh positive if he has the Rh antigen, and Rh negative otherwise. He is AB, A, or B depending on which of the other antigens he has, with type O having neither A nor B. Draw a Venn diagram and identify each of the eight areas.

14. [From Kemeny et al. (1966)] Considering only the two subsets, the set X of people having antigen A and the set Y of people having antigen B, define (symbolically) the types AB, A, B, and O.

15. Suppose that there are four political parties in a legislature: A, B, C, and D. Suppose that no party can win a vote on its own, but that the largest party A plus any one of the smaller parties B, C, and D can win or the three smaller parties can win if they form a coalition.

   (a) List the power set for the set of political parties {A, B, C, D}.
   (b) Divide the power set into two subsets, the set W of *winning* coalitions and the set L = W' of *losing* coalitions.

16. Using the simulation *Venn Diagrams*, can you create a Venn diagram for three sets, represented by circles, that intersect pairwise: Circles A and B, B and C, and A and C intersect but $A \cap B \cap C = \emptyset$?

17. Prove that $(A \cap B)' = A' \cup B'$ and $(A \cup B)' = A' \cap B'$.

CHAPTER 3

Probability: Pure and Applied

We all have an intuitive understanding of probability. We use it in everyday language, when we say that it probably won't rain tomorrow. We also use probabilities more precisely when we say that the probability that a fair coin will show heads when flipped is .50, or the odds are 35 to 1 against two rolled dice both showing ones ("snake eyes"). In this chapter we will explore a more explicit mathematical definition of probability that will enable us to calculate probabilities in more complex situations. Probability is the basis for statistics, and we will learn something about how to test hypotheses using statistics.

The first thing we will look at is a precise definition of probability. How do we know a probability when we see one? For example, we know that no probability is negative or greater than 1.00. Do they have other essential characteristics? We want a sufficient set of criteria for probabilities. Note that this is a different question from whether a probability is accurate. For example, if a die is warped its probability of showing a 1 may be something other than the 1/6 it would be if the die were fair.

We start out with the set S of all the possible outcomes of some process. Here are some examples:

1. We flip a coin. $S = \{\text{heads, tails}\}$
2. We roll a die. $S = \{1, 2, 3, 4, 5, 6\}$
3. We sample one person from a classroom of six students. $S = \{\text{Yamile, Daphne, Stephanae, Curtis, Jennifer, George}\}$
4. We can draw one card from an ordinary deck of 52 cards. $S = \{2 \text{ of clubs, } 3 \text{ of clubs, } \ldots, \text{ ace of spades}\}$

These basic outcomes are known as *elementary events*, to distinguish them from larger events, called simply *events*. For example, consider rolling a die and getting an even number. This event $E = \{2, 4, 6\}$. Rolling an odd number is $O = \{1, 3, 5\}$. Note that E and O are both subsets of $S : E \subseteq S$ and $O \subseteq S$. Rolling a negative number cannot happen, so that event $= \varnothing$. Every event is a subset of the set of elementary events and a member of the power set 2^S. With the die, $E \in 2^S$ and $O \in 2^S$.

We are now in a position to define a probability function. A probability function is an assignment of numbers $P(A)$ to all the events A in 2^S, the power set for all the elementary events, with the following three properties:

1. $P(A) \geq 0$ for every event A.
2. $P(S) = 1$
3. If there exists some countable set of events $A_1, A_2, \ldots$ and if these events are all mutually exclusive, then $P(A_1 \cup A_2 \cup \ldots) = P(A_1) + P(A_2) + \ldots$

The first rule says that no probability can be negative, the second says that the probability that something will happen is one, and the third says that the probability that any one of the number of mutually exclusive events will occur is the sum of their separate probabilities. From these basic assumptions every other characteristic of probabilities can be demonstrated. For example, you can show that $P(\varnothing) = 0$.

Let's use these rules to calculate the probabilities of drawing various types of cards from a deck.

1. **What is the probability of drawing the ace of spades from a well-shuffled deck?** If the deck is well shuffled, then each of the elementary events has the same probability p. The elementary events are mutually exclusive and together constitute S, the set of all elementary events. Therefore $p + p + \ldots + p = 1$ (p 52 times), using Properties 2 and 3 of probabilities. Therefore, $p = 1/52$.

2. **What is the probability of drawing a king?** There are four mutually exclusive elementary events corresponding to drawing a king: the kings of hearts, diamonds, and clubs, and spades. Each of them has a probability of $1/52$. By Property 3, the probability of drawing a king is $1/52 + 1/52 + 1/52 + 1/52 = 4/52$. Notice that the probability of an event equals the sum of the probabilities of the elementary events it contains (because elementary events are mutually exclusive) and that if all the elementary events are equally likely to occur then the probability of an event is just the number of elementary events constituting the event divided by the number of elementary events.

3. **What is the probability of drawing a king or an ace?** There are 8 cards that are kings or aces. Therefore the probability of drawing a king or an ace is $8/52$.

4. **What is the probability of drawing an ace or a spade?** There are 13 spades and 4 aces, but there are only 16 (not 17) cards that are either an ace or a spade. Therefore, the probability is $16/52$.

This last result can be expressed by a more general formula. Suppose that A and B are two events. If $A \cap B = \varnothing$ then $P(A \cup B) = P(A) + P(B)$, by Property 3 of probability functions. However, what if A and B are not mutually exclusive, as in the last example?

Then,

$P(A) = P(A \cap B) + P(A \cap B')$ by Property 3
$P(B) = P(B \cap A) + P(B \cap A')$ by Property 3

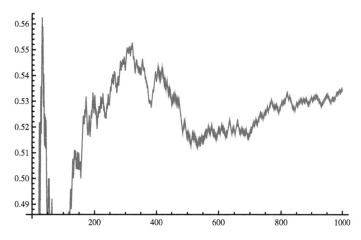

Figure 3.1. The convergence of proportions of heads to probabilities.

$P(A) + P(B) = P(A \cap B) + P(A \cap B') + P(B \cap A) + P(B \cap A') = P(A \cup B) + P(A \cap B)$ by Property 3

Therefore, $P(A \cup B) = P(A) + P(B) - P(A \cap B)$

It is easy to show that the probability of a set plus the probability of its complement equals 1.

$P(A) + P(A') = P(S) = 1$ by Properties 2 and 3

Therfore, $P(A') = 1 - P(A)$

Thus far we have defined an abstract probability function, but we have not described how probabilities connect to the real world. What is the probability that a flipped penny will show heads? It need not be exactly .50 because the penny is not absolutely symmetric and it may become worn and warped with time. The following expresses the relation between probabilities and the real world and shows how we can estimate probabilities.

If the probability of occurrence of an event X is $P(X)$ and N trials are made, independently and under exactly the same circumstances, then the relative frequency of the event will approach the probability of the event.

Let me give you an example. I let the computer flip a simulated coin 1,000 times. This was a biased coin whose probability of heads was .55. Figure 3.1 shows the cumulative proportion of heads through these 1,000 simulated flips.

The demonstration *Convergence* enables you to examine these ideas for yourself. The demonstration enables you to set the probability of heads (p) and the number of flips (N). Two charts are presented: the proportion of heads as the number of trials t increases and the difference between the number of heads and the expected number of heads, $t \times p$. The charts show that the proportion of heads gets closer and closer to p but, a bit

Table 3.1.

Payoffs for roulette bets

Bet	Payoff for a win	Probability of winning
Red or black	$1	18/38
Single number	$35	1/38
Split bet (two adjacent numbers)	$17	2/38
Corner bet (four adjacent numbers)	$8	4/38

counterintuitively, the number of hears does not approach $N \times p$. Instead the difference tends to increase with the square root of the number of trials.

EXAMPLE: GAMBLING

Let's apply these ideas to gambling on a roulette wheel. In Nevada roulette wheels have 38 pockets: 18 red, 18 black, and 2 green. The 36 red and black pockets are numbered from 1 to 36 while the 2 green pockets are labeled "0" and "00." There are a great number of different bets that can be placed. Table 3.1 shows some of the bets that can be made and their payoffs for a dollar bet.

Suppose you place 1,000 bets on red (or black). By the principle that proportions get closer and closer to probabilities, you should expect to win 18/38 of the time, or about 474 times and lose about 526 times. Thus you should expect to lose around $52, or $.052 per dollar bet.

To calculate the average loss we did the following calculation:

$$\frac{\left(1000 \times \frac{18}{38} \times \$1\right) + \left(1000 \times \frac{20}{38} \times (-1\$)\right)}{1000} \tag{3.1}$$

Note that the number of times we bet, 1,000, cancels out of the numerator and denominator and we are left with the more general formula: $\frac{18}{38}(\$1) + \frac{20}{38}(-\$1) =$ probability of winning $\times$ profit amount + probability of loss $\times$ loss amount $= -\$.052$.

This is called the *expected value* of the bet. That the expected value is negative means that, on the average and in the long run, you will lose (subject to random fluctuations) 5 cents out of every dollar you bet.

Hoping to improve our expectations, let's look at the expected value of a bet on a single number.

$$\frac{1}{38}(\$35) + \frac{37}{38}(-\$1) = -\$.052 \tag{3.2}$$

Alas, the expected loss is the same. This is not a better bet.

The evils of gambling and the concept of expected value are illustrated in the demonstration *Dice Gambling*. In the initial setting the dice game is fair (it's expected value is zero) except for the fee of $1 per game. This means that the total expected value is −1. If you play the game enough times you will see that you will lose all your money more frequently than you will double your money, but if you set the fee to zero you will double your money as frequently as you lose it all.

The demonstration *Dice Gambling* illustrates the folly of gambling games of chance against a gambling establishment. The game itself is made up to be especially transparent: it is not an actual Las Vegas game. A *fair* game is one in which the expected gain to the gambler is zero. All commercial gambling houses guarantee that the expected gain to the gambler is negative. This can be accomplished in two ways. First, the underlying game may be fair but the gambler may be charged a fixed amount for every time he plays. Second, the expected gain can be negative. The first pattern is illustrated with the default game. The gambler wins $30 if the die comes up 4, 5, or 6 and loses $30 if it comes up 1, 2, or 3. This would be a fair game (its expected value is zero) except that the gambler is charged $1 every time she plays. The gambler, who starts with $150, plays until she doubles her money or goes broke. You can easily show that the gambler loses all her money more often than she doubles it.

The second house strategy is illustrated by setting the fee to zero but making the underlying game unfair, for example, modify the game so that when the player wins he earns $25 but he loses $30. The same pattern will emerge; the gambler will lose his money more frequently than he will reach his goal.

TWO OR MORE EVENTS: CONDITIONAL PROBABILITIES

Let me start with an easy card problem. What is the probability of drawing two aces from a deck of cards? The sample space consists of all 52×51 sequences of two cards drawn (without replacement) from a deck of 52 cards. Let's let A_1 be the event that the first card is an ace; this can happen in 4×51 ways and $P(A_1) = 4 \times 51/(52 \times 51)$. A_2 is the event that the second card is an ace; this can happen in 51×4 ways and $P(A_2) = 51 \times 4/(52 \times 51)$. "$A_2|A_1$" is the event that second card is an ace if the first card is an ace; it involves a restriction of the sample space to the deck of 51 cards with 3 aces if the first card drawn was an ace, and $P(A_2|A_1)$ be the probability of drawing a second ace if a first ace has already been drawn. $P(A_2|A_1) = 3/51$. The probability of drawing two aces is the proportion of sequences in the sample space with two aces; $P(A_1 \cap A_2) = 4 \times 3/(52 \times 51)$.

If you look at these numbers you will find the following to be true:

$$P(A_1 \cap A_2) = P(A_1)P(A_2 \mid A_1) \tag{3.3}$$

and,

$$P(A_1) \neq 0, \ P(A_2 \mid A_1) = \frac{P(A_1 \cap A_2)}{P(A_1)} \tag{3.4}$$

Table 3.2.

Movie experience by sex

	X-rated movie (X)	Non-x-rated movie (X)'	Total
Male (M)	137	294	431
Female (F)	96	468	563
Total	233	762	995

Is this a coincidence? Of course not! The equation simply says that if two events are to occur the first must occur and then the second one must occur within the reduced sample space of events in which the first has occurred.

Now let's look at a more sociological example. In a national random sample men and women were asked whether or not they had seen an X-rated movie in the past year. The results are given in Table 3.2.

Suppose we sample one person at random from these 995. We can calculate various informative conditional probabilities, such as the probability that a man or a woman has seen an X-rated picture in the past year.

$$P(X \mid M) = \frac{137}{431} = .318 \tag{3.5}$$

$$P(X \mid F) = \frac{96}{564} = .170 \tag{3.6}$$

There are a number of ways we can calculate the probability that the person sampled will be both a male and have recently seen as X-rated picture. First, we can simply calculate the ratio of the number of elementary events in $M \cap X$ to the total number of elementary events.

$$P(M \cap X) = \frac{133}{995} \tag{3.7}$$

Alternatively, there are two different ways to use conditional probabilities.

$$= (431/995)(137/431) = 137/995$$
$$P(M \cap X) = P(M)P(X \mid M) = \left(\frac{431}{995}\right)\left(\frac{137}{431}\right) = \frac{137}{995} \tag{3.8}$$

$$P(M \cap X) = P(X)P(X \mid M) = \left(\frac{233}{995}\right)\left(\frac{137}{233}\right) = \frac{137}{995} \tag{3.9}$$

TWO OR MORE EVENTS: INDEPENDENCE

In some of the examples above the probability of an Event B is unaffected by whether or not Event A has occurred, while for other examples this is not the case.

1. Whether a flipped coin is heads or tails on a second toss is unaffected by whether or not it was heads on the first.
2. When drawing without replacement, the probability of a second card being an ace depends on whether the first card was an ace.
3. If two cards are drawn from a deck with replacement of the first card before the second is drawn, then the probability of a second ace is not affected by whether the first card was an ace.
4. The probability that someone has recently seen an X-rated movie is affected by whether that person is male or female.

If the probability of B is not affected by whether A has occurred we say that A and B are *independent*.

Definition. If $P(B \mid A) = p(B)$, then A and B are said to be independent

Note that if A and B are independent, the $p(A \cap B)$ has a particularly simple form: the probability of the intersection is the product of the probabilities of the separate events.

If A and B are independent, $P(A \cap B) = P(A \mid B)P(B) = P(A)P(B)$.

Let me take another example. Suppose that the probability that a patient with a particular disease improves if he is given a new treatment is .50. Suppose that six patients are randomly sampled and given the new treatment. What is the probability that all improve? Let A_i be the event that the ith sampled patient improves. Using the multiplication rule for independent events,

$$P(A_1 \cap A_2 \cap A_3 \cap A_4 \cap A_5 \cap A_6) = \qquad (3.10)$$

$$P(A_1) \times P(A_2) \times P(A_3) \times P(A_4) \times P(A_5) \times P(A_6) = (1/2)^6$$

A COUNTING RULE: PERMUTATIONS AND COMBINATIONS

There is one other bit of mathematics you must know—permutations and combinations. A permutation is a rearrangement of a set of objects. Suppose, for example, that you are holding three cards in your hand: an ace (A), king (K), and queen (Q). There are six orders of these cards, from left to right: AKQ, AQK, KAQ, KQA, QAK, QKA. The number of ways that three objects can be arranged is $3 \times 2 \times 1 = 6$. If you had five cards in your hand, the number of ways they could be arranged is $5 \times 4 \times 3 \times 2 \times 1 = 120$. There are $52 \times 51 \times \ldots \times 1 = 80, 658, 175, 170, 943, 878, 571, 660, 636, 856, 403, 766, 975, 289, 505, 440, 883, 277, 824, 000, 000, 000, 000$ ways that a full set of 52 cards can be ordered. In general, n objects can be ordered in $n \times n - 1 \times \ldots 1 = n!$ different ways. $n!$ is read as "n factorial."

A related question concerns how many ways m objects can be selected from n objects regardless of order. For example, how many poker hands are there if 5 cards are drawn from an ordinary deck of 52 cards and the order in which the 5 cards is drawn is ignored. Or, suppose that a well-meaning teacher with 25 students wants to appoint a different 5-person student council in her class every day. How long will it take her to appoint all the possible councils? The answer is given by the following expression, where the left side is read "binomial coefficient n over m".

$$\binom{n}{m} = \frac{n!}{m!(n-m)!} \tag{3.11}$$

So, there are $52!/(5!47!) = 2{,}598{,}960$ different poker hands, and there are $25!/(5!20!) = 53{,}130$ different ways in which 5 students can be selected from 25 students. If the teacher appointed 5 councils a week for 40 weeks every year it would take her more than 265 years.

Let's use the binomial coefficient to calculate the probabilities of some poker hands. A poker hand is any 5 cards from an ordinary deck of 52 cards. What is the probability of being dealt a flush, 5 cards all of the same suit? A flush involves selecting 5 cards from 13 cards, the number of cards in a suit. There are four suits. The probability of drawing a flush is the number of flushes divided the total number of poker hands.

$$\frac{4\binom{13}{5}}{\binom{52}{5}} = \frac{33}{1660} = .002 \tag{3.12}$$

THE BINOMIAL DISTRIBUTION

Finally we have all the tools we need to talk about a simple use of statistics to test hypotheses in sociology or any other field. Let's go back to an earlier example. Suppose six randomly selected patients with a disease were matched with six other very similar patients. In other words, Harry, a severely ill 25-year-old man, was matched with George, a 26-year-old very sick man, while Harriet, a 40-year-old moderately afflicted woman, was matched with Georgiana, an equally sick 40-year-old woman. Within each pair of equally sick patients one was randomly chosen to receive the drug while the other received a placebo. At the end of the experiment all patients were evaluated and within each pair it was determined whether the patient receiving the new drug or the patient receiving the placebo was better off. Each instance in which the patient receiving the new drug is healthier is a *success* with respect to the new drug. If the drug is in fact no better than the placebo the experimental patient will be no more likely than the placebo-receiving patient to do better; the probability that the drug-receiving patient turns out to be the healthier of the two should be .50 for each pair. If the drug is more effective, then this probability should be more than .50 that the experimental patient in each pair does better.

Table 3.3.
All sequences of four successes and two failures in six trials and their probabilities

Sequence	Probability	Sequence	Probability
FFSSSS	$qqpppp$	SFSSSF	$pqpppq$
FSFSSS	$qpqppp$	SSFFSS	$ppqqpp$
FSSFSS	$qppqpp$	SSFSFS	$ppqpqp$
FSSSFS	$qpppqp$	SSFSSF	$ppqppq$
FSSSSF	$qppppq$	SSSFFS	$pppqqp$
SFFSSS	$pqqppp$	SSSFSF	$pppqpq$
SFSFSS	$pqpqpp$	SSSSFF	$ppppqq$
SFSSFS	$pqppqp$		

Let p be the probability that the experimental patient is healthier at the end of the experiment while $1 - p = q$ is the probability that the placebo-receiving drug is healthier. The probability that there are six successes is p^6. What are the probabilities of five or four or three successes? To calculate this we will break the problem down into two parts: what is the probability of each sequence of x successes and $6 - x$ failures, and how many sequences of x successes are there.

For example, Table 3.3 shows every sequence in which there are four successes and two failures. The probabilities are calculated using the fact that results for the pairs of patients are independent.

Looking at the second column of probabilities, every one of them has 4 ps and 2 qs. All of these probabilities are equal to p^4q^2. More generally, if there are m successes and $n - m$ failures in n trials in a sequence, the probability of that particular sequence is $p^m q^{n-m}$.

There are 15 sequences of 2 failures and 4 successes because there or $6!/(4!2!)$ ways of selecting the 4 pairs (out of 6) with the successes. More generally, the number of ways of selecting the positions of m successes with n trials is the binomial coefficient n over m, represented by Equation 3.11.

Putting all this together, we know how many sequences there will be with m successes in n trials and we know the probability of each sequence, $p^m q^{n-m}$. Therefore we have the formula for the probability of m successes in n trials where the probability of a success in any trial is $p, bP(n, p, m)$.

$$P(n, p, m) = \binom{n}{m} p^m q^{n-m} \qquad (3.13)$$

Table 3.4 shows the probabilities of various numbers of successes in six trials if the probability of a success in any one trials is .80 (the treatment is much better than a placebo). The right column represents when the new drug is no better than the placebo.

There are two demonstrations that are designed to familiarize you with the binomial distribution. *Binomial Trials* generates the number

Table 3.4.
Probabilities of successes of an effective versus ineffective drug

Number of successes	Effective drug probability	Ineffective drug probability
0	$\frac{6!}{0!6!} \times .8^0 \times .2^6 = 0$	$\frac{6!}{0!6!} \times .5^0 \times .5^6 = 0.016$
1	$\frac{6!}{1!5!} \times .8^1 \times .2^5 = 0.002$	$\frac{6!}{1!5!} \times .5^1 \times .5 = 0.094$
2	$\frac{6!}{2!4!} \times .8^2 \times .2^4 = 0.15$	$\frac{6!}{2!4!} \times .5^2 \times .5^4 = 0.234$
3	$\frac{6!}{3!3!} \times .8^3 \times .2^3 = 0.082$	$\frac{6!}{3!3!} \times .5^3 \times .5^3 = 0.312$
4	$\frac{6!}{4!2!} \times .8^4 \times .2^2 = 0.246$	$\frac{6!}{4!2!} \times .5^4 \times .5^2 = 0.234$
5	$\frac{6!}{5!1!} \times .8^5 \times .2^1 = 0.393$	$\frac{6!}{5!1!} \times .5^5 \times .5^1 = 0.094$
6	$\frac{6!}{6!0!} \times .8^6 \times .2^0 = 0.262$	$\frac{6!}{6!0!} \times .5^6 \times .5^0 = 0.016$

of successes in 50 binomial experiments in which you can select the number of trials per experiment and the probability of a success in each independent trial. After 50 trials the histogram will be approximately the shape of the binomial distribution for that combination of n and p.

Binomial Fit is a game in which you are asked to guess the n and p for an arbitrary set of binomial experiments. You should vary the n and p sliders until the curve approximates the distribution as closely as possible. The Evaluate button will show you the correct answer.

Now let us decide how the experiment should turn out before we have some faith in the new drug, enough to mount a more expensive study with a larger sample. We can make two kinds of mistakes and two types of correct decisions, depending on what we conclude and whether or not the drug is effective, $p > .50$.

Correct Decision 1: Drug is ineffective and our experiment shows it is ineffective.

Correct Decision 2: The drug is effective and we conclude it is effective.

Type I Error: Our experiment decides an ineffective drug is effective.

Type II Error: Our experiment decides an effective drug is ineffective.

Types I and II errors have different costs. A Type II error means that a possibly useful drug is missed. A Type I error means that an expensive study is mounted to test an ineffective drug. How much protection does this study offer against Type I and Type II errors? The probability of a Type I error, usually labeled α, depends on the number of successes we require before we classify the drug as worthy of further study. Suppose we are conservative and decide that the experimental patient in every pair must do better than the control subject before we decide that the new medicine is promising. A type I error occurs if an ineffective drug passes this test. From the right column of Table 3.4 we can see that the probability that an ineffective drug will pass this hurdle is very small, .016.

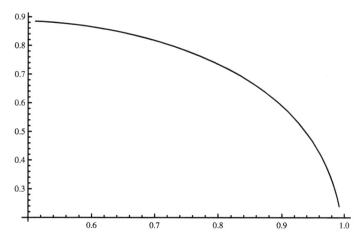

Figure 3.2. Probabilities of Type II error.

The probability of committing a Type II error, called β, depends on how good the drug is. The right column of Table 3.4 shows that even if the experimental patient is four times as likely to improve more than his placebo partner, the probability that the new drug will pass our test is just .262. The probability of a Type II error is therefore .738 in this situation. Figure 3.2 plots the true probability that the experimental patient will do better than the control patient versus β, the probability of a Type II error. Figure 3.2 shows that our study is likely to misclassify all but the most potent drugs.

Of course, lowering the criterion from six to five successes would reduce the probabilities that effective drugs would be misclassified but only at the expense of a greater probability of accepting ineffective drugs (Type I error). One way to reduce both α and β would be to increase the sample size from 6.

The demonstration *Binomial Power* allows you to examine the trade-off between committing Type I and Type II errors and how increasing the number of trials (number of matched pairs of patients) can reduce the probabilities of both types of error. When you start this demonstration the number of trials is 10 and the probability of a Type I error that you have selected is .05. The chart shows the power of the test against various probabilities that the new drug is better than the placebo. The power of a test is $1 - \beta$, the probability that a better drug will be correctly detected by the test. By playing with the sliders you can see that the power of the test increases with the number of trials in the experiment.

Chapter Demonstrations

- *Dice Gambling*
- *Binomial Trials*

- *Binomial Fit*
- *Convergence*
- *Binomial Power*

EXERCISES

1. Two dice, one red and the other green, are rolled.

    (a) How many elementary events (outcomes) are there?
    (b) In how many of these elementary events are the faces of the two dice equal?
    (c) What is the probability that both dice will have the same value?
    (d) What is the probability of rolling a 7?
    (e) What is the probability that at least one of the dice shows a 6?
    (f) If you rolled these dice 1,200 times, how many times would you expect the two dice to be equal?
    (g) Suppose someone offers to bet you $10 that two dice you roll will not be equal. If they are equal he pays you $10 and if they are not equal you pay him $2. Is this a fair bet?

2. Suppose that you draw 5 cards without replacement from an ordinary deck of 52 cards.

    (a) What is the probability that the first two cards are clubs?
    (b) What is the probability that the first two cards have the same suit?
    (c) What is the probability that the first two cards are from different suits?
    (d) What is the probability that the first three cards are all from different suits?
    (e) What is the probability that all five cards are from different suits?
    (f) What is the probability of drawing a straight flush—five consecutively valued cards from the same suit?
    (g) What is the probability of drawing a straight—cards with five consecutive values?

3. Examine Table 3.2, the cross-classification of sex and seeing X-rated films. What would the cells of the table look like if sex and seeing X-rated movies were independent? Suppose that the numbers of men, women, people recently seeing X-rated films, and people not have not changed.

4. Use the simulation *Convergence* to answer the following questions.

    (a) Set $p = .5$, representing the flipping of a fair coin. Does the surplus of heads decrease over time?
    (b) Does the proportion of heads get closer and closer to .50?

5. Use the simulation *Binomial Power* to answer the following questions. You are interested in the relative performance of students in charter schools versus ordinary public schools. To begin your study you locate

10 communities with both charter and public schools. If there is preliminary evidence that the charter schools are performing better you will begin a full-scale study of the differences.

(a) You decide that if 8 or more of the charter schools must have higher performing students than their comparable public schools before you are willing to begin the larger study. What is a Type I error? What are its implications? What is α?

(b) What are the implications of a Type II error in this case? If the charter schools are twice as likely to perform better, what is the power of the test?

(c) If you wanted the probability of a Type I error to be only 1%, how many charter schools would have to perform better before you would launch the larger study?

(d) If you wanted the probability of a Type I error to be .05, and you wanted the probability of detecting the superiority of the charter schools if they are better two-thirds of the time to be .80, how big should your sample be?

CHAPTER 4

Relations and Functions

Sociology's special mission is to analyze patterns of relations between people in groups. This is what distinguishes us from psychologists, who study what goes on inside the heads of individuals. Table 4.1 on the next page gives some typical sociology courses.

Of course other social sciences also specialize in studying patterned and organized activities of human groups; economics, political science, and anthropology all study relations between people. There is, however, one field that studies relations in the abstract, regardless of the objects the relations connect. Mathematics has a special and very well-developed language for describing relations of all sorts. This language has proved to be incredibly useful to social scientists who study relations between people. So, we will now plunge into some mathematical terminology, coming to the surface occasionally to catch our breath and make the connections to the social sciences.

The mathematical definition of a relation begins with a set. This set could be anything at all. The set could be the numbers $\{1, 2, 3, 4\}$ or the people {Andy, Barbara, Carl}. Next we consider all the *ordered pairs* of these objects.

Definition. Let S be a set. The set $S \times S$, the *Cartesian product* of the set S with itself, consists of all ordered pairs (a, b) where $a \in S$ and $b \in S$.

The order makes a difference: $(a, b) \neq (b, a)$. Also, do not confuse (a, b), the ordered pair, with $\{a, b\}$, the set consisting of a and b. $\{a, b\} = \{b, a\}$ because the order in which the elements in the definition of a set is written is irrelevant. Also, (a, a) as an ordered pair is included in the definition of an ordered pair, but $\{a, a\}$ makes no sense.

Table 4.2 gives all the ordered pairs of the set consisting of Andy (A), Barbara (B), and Carl (C).

For the set N consisting of the numbers 1, 2, 3, and 4, there would be $4 \times 4 = 16$ ordered pairs in $N \times N$. If S consists of n elements, then there are n^2 ordered pairs of elements in the set $S \times S$.

Definition. A *dyadic relation* R is a subset of the set $S \times S$, written as $R \subseteq S \times S$.

Table 4.1.

Typical sociology courses

Course	Description
Sociology of the family	Such a course will deal with the family and its role of the family in the larger society. For example, the family socializes children, transmits social class, provides intimacy, and performs other important activities.
Political sociology	Political sociology will cover social movements that advocate change, political parties, the effects of organized interest groups, and social classes.
Urban sociology	Urban sociology courses will describe the forms of city governments, the roles of political parties and machines, and organized interest groups.
Formal Organizations	This course will analyze how firms, bureaucracies, and other highly organized groups function and relate to one another.
Sociology of education	This course will describe the US educational system, the organization of school systems, and the social organization of teachers and students within schools.
Social psychology	An important part of any social psychology course taught in a sociology department will be about group dynamics, the processes, such as leadership, that occur in small groups.
Deviance	Such a course will show how deviance from the larger society's norms occurs in groups. The course will cover adolescent street gangs, criminal adult groups, and deviant subcultures.
Race relations	Such a course will likely cover the internal organization of minority communities.

Table 4.2.

All possible ordered pairs with A, B, and C

(A,A)	(A,B)	(A,C)
(B,A)	(B,B)	(B,C)
(C,A)	(C,B)	(C,C)

For example, let's look at some relations involving $N = \{1, 2, 3, 4\}$. Consider the relation "greater than," which a mathematician writes as " $>$ ", as in "3 $>$ 2." This relation, call it G, consists of the following subset of the elements of $N \times N$.

$$G = \{(2, 1), (3, 1), (3, 2), (4, 1), (4, 2), (4, 3)\} \subseteq N \times N \qquad (4.1)$$

Now consider the relation "greater than or equal to," GE for short, which a mathematician would write as "$\geq$." This consists of a different subset of $N \times N$:

$$GE = \{(1, 1), (2, 1), (2, 2), (3, 1), (3, 2), (3, 3), (4, 1), (4, 2), (4, 3), (4, 4)\}$$
$$\subseteq N \times N \qquad (4.2)$$

Next, consider the relation "divides evenly, without a remainder," whose symbol is "|." This relation consists of the following ordered pairs of N.

$$\text{"} | \text{"} = \{(1, 1), (1, 2), (1, 3), (1, 4)(2, 2)(2, 4)(3, 3)(4, 4)\} \subseteq N \times N \qquad (4.3)$$

Now consider a relation among the set of three people, $P = \{$Andy, Barbara, Carl$\}$. Suppose that Andy likes Barbara, Barbara likes Andy, and Carl likes Barbara. Let's call this relation L. Then L consists of the following subset of $P \times P$.

$$L = \{(A, B), (B, A), (C, B)\} \qquad (4.4)$$

There is an alternative way of expressing a dyadic relation. Instead of saying that "(x, y) is one of the ordered pairs contained in a relation R," we write "xRy." For example, $3 > 2$, $2 \mid 4$, and Andy L Barbara (Andy likes Barbara).

Relations can also be represented by diagrams in which the elements of the set are points and relations are shown by lines between points. Arrows on the lines show the direction of the relation. aRb is represented by $a \to b$ in the diagram. Diagrams of this sort are called *graphs*. The graph of the relation $>$ among the numbers 1, 2, 3, and 4 is given in Figure 4.1.

As another example, consider the relation "respect" among four people, Ricky Ricardo, Lucy Ricardo, and their landlords and friends, Fred Mertz and Ethel Mertz. Having watched all episodes of this early TV hit repeatedly, I feel I can say confidently that everyone respects Ricky and that Ricky and Fred always show respect for one another, but that no other pair show any degree of respect or deference. Figure 4.2 diagrams these relations.

Every sociologist knows that some characteristics of people are basic and important, no matter what one is studying: economic behavior, political attitudes, sexual behavior, or marriage partner choice. Some of these characteristics are social class, gender, race and ethnicity, nationality, and age. If you know someone's age, sex, income, race, and citizenship, you know one heck of a lot about her. Similarly, there are some basic but very abstract properties of the relationships that form a network that are important no matter what the content of the relationships, and it is to these we will now turn.

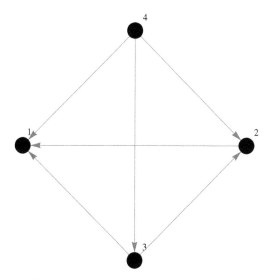

Figure 4.1. Liking relations among people and "greater than" relations among numbers.

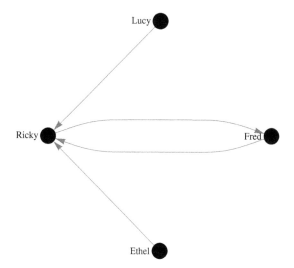

Figure 4.2. Respect among the Ricardos and Mertzs.

SYMMETRY

Definition. *R* is a *symmetric* relation if whenever $(a, b) \in R$, then $(b, a) \in R$ also. In other words, aRb implies bRa.

The "respect" relation among Ricky, Lucy, Fred, and Ethel in Figure 4.2 is not symmetric. Lucy respects Ricky, but Ricky does not respect Lucy. It's true that Ricky and Fred respect each other, but for a relation to be

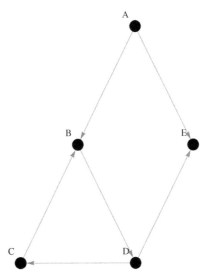

Figure 4.3. An antisymmetric relation.

symmetric, every single relationship must be reciprocated. In the graph of a symmetric relation, there are no pairs connected by just one line with an arrow. Some relationships among people are perfectly symmetric: "is married to," "is the sibling of," "has sexual intercourse with." Others are symmetric in a statistical sense. For example, if you know that *A* likes *B*, then the odds are that *B* likes *A*. The same holds for dislike.

Relations that are usually symmetric can have dramatic consequences when they are not. We all know the tragic or comic consequences that can occur when love is not reciprocated, but even more mundane relations can be interesting when not symmetric. There is a common rule of social life, sometimes called the *norm of reciprocity* by sociologists, that peers ought to reciprocate favors. Someone who continually lets others pay for his restaurant meals develops an undesirable reputation as a freeloader. There is likely to be an internal sense of discomfort if one is indebted to another unless the other has more status or power. Children accept their dependence on their parents, and mentees on their mentors, but the cost of being dependent is inequality of power. To be indebted to a peer is to risk losing status and power in a previously equal relationship.

Definition. A relationship *R* is *antisymmetric* if whenever *aRb* and *bRA*, then *a = b*.

Another equivalent way of putting this is that a relationship is antisymmetric if for every pair *a* and *b*, *a ≠ b*, either there is no relationship, or *aRb*, or *bRa*, but it is never true that *aRb* and *bRa*. In the graph of an antisymmetric relation, there are no pairs connected by lines with arrows running both ways. Consider, for example, the relationship in Figure 4.3.

Figure 4.3 represents an antisymmetric relation because there are no two-way relationships. In mathematics, the relations $\geq$ and $>$ are antisymmetric. In an organization, the relation "is the superior of" is antisymmetric. Among chickens, "pecks" is antisymmetric because if one chicken pecks the other she is not pecked back in return. Among the basketball teams in a tournament, the relation "beats" is antisymmetric. If no pair of teams plays more than one game against each other, then either they don't play, or a beats b or b beats a; these are the only possible outcomes.

Symmetry and antisymmetry are opposites. If a relationship is symmetric, then all relationships between different objects (people or numbers or whatever) are two-way, but for antisymmetric relations all relationships are one-way.

Let's look at some of the mathematical relations and ask whether they are symmetric, nonsymmetric, or antisymmetric.

1. This relation $>$ is antisymmetric. If $a > b$, then it is never true that $b > a$. Just look at the list of pairs defining this relation. $(2, 1)$ is in the relation, but $(1, 2)$ is not. $(3, 2)$ is in the relation, but $(2, 3)$ is not.
2. The $\geq$ relation is also antisymmetric. The only pairs for which (a, b) and (b, a) are both in the relation are those for which $a = b$: $(1, 1)$, $(2, 2)$, $(3, 3)$, and $(4, 4)$.
3. The equality relation is clearly symmetric.
4. The $|$ relation is also antisymmetric. Note that two numbers can divide each other evenly only if they are equal.

The liking relation among Andy, Barbara, and Carl is not symmetric. The reason is that (Carl, Barbara) is in L, but (Barbara, Carl) is not. On the other hand, it is not antisymmetric because Andy and Barbara (different people) like each other.

This "respect" relation among the Ricardo and Mertz families also is not symmetric or antisymmetric. Lucy and Ethel appear to respect and defer to Ricky but not visa versa. But, Fred and Ricky respect each other.

REFLEXIVITY

Definition. A relationship is *reflexive* if (a, a) is in the relation for all a in S (or, equivalently, aRa for all a in S).

Whereas symmetry is defined by the relations between pairs, reflexivity is a relationship of the individual to itself. We have discussed a number of mathematical relations that are reflexive: "$=$," "$|$," and "$\geq$." Neither the liking relations among Andy, Barbara, and Carl (Figure 4.1) or the respect relations among Ethel, Lucy, Ricky, and Fred (Figure 4.2) is reflexive. In fact, when you think about it, self-liking and self-respect refer to completely different phenomena than liking and respect for another person. We, the authors, cannot think of a single significant relation between people that is also reflexive. However, there are meaningful relations between groups of people that may or may not be reflexive. For example, most university departments have pretty strong norms prohibiting hiring their own recent graduate students as assistant professors, but business firms recruit their top managers from their own ranks.

TRANSITIVITY

A more subtle property of relations described by mathematicians is *transitivity*. Symmetry involves two objects and reflexivity involves just one, but the definition of transitivity involves three objects from the set S (some of them may be repeated). A relationship R is transitive if whenever (a, b) and (b, c) is in the relation, so is (a, c). It's obvious that the relations ">" and "=" are transitive among numbers.

It is also easy to see to see that the relation L among Andy, Barbara, and Carl is not transitive. (Carl, Barbara) and (Barbara, Andy) are both in L, but (Carl, Andy) is not. You should also be able to show that the "respect" relation among Ricky, Fred, Ethel, and Lucy is also not transitive. The relation in Figure 4.3 is not transitive (Do you see why?). However, there are human relations that are transitive. If a is an ancestor of b and b of c then a is also an ancestor of c. If a and b have the same religion and b and c have the same religion, then a and c have the same religion. For a transitive relation expressed as a graph, you never see a situation in which $a \rightarrow b$, $b \rightarrow c$, but no $a \rightarrow c$.

Figure 4.4 shows all the possible types of patterns among three people. Some of them are transitive and others are not.

Which of these patterns are transitive? Some of them, like 4 and 13, have intransitive triples that clearly violate transitivity: in 13 the person in the lower right chooses the person in the lower left who, in turn, chooses the top person, but the person in the lower right does not choose the top person, violating transitivity. Other patterns, like 7 and 11, contain one or more triplets that illustrate transitivity and none that violate it. Finally, there are a number of networks (1 and 2 for example) that contain neither triplets contradicting transitivity nor triplets illustrating it.

Definition. A network is *transitive* if and only if it contains no intransitive triplets: no instance of $(a, b) \in R$, $(b, c) \in R$, but $(a, c) \notin R$.

Note that by this definition all the networks in the third category, neither contradicting nor illustrating transitivity, are transitive. The following are the transitive networks: 1, 2, 3, 5, 6, 7, 11, 12, 16. The same definition applies to all networks, regardless of their size: a network is transitive only if it does not contain an intransitive triplet.

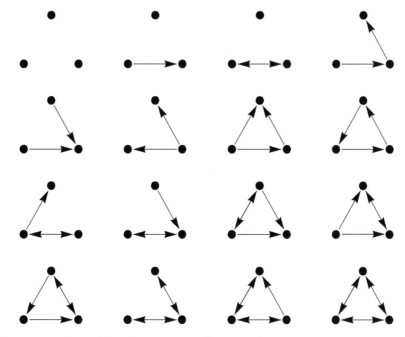

Figure 4.4. All possible relations among three people.

The very existence of human society depends on the fact that humans can form transitive relations because transitive relations allow for consistent hierarchies of great depth. Large organizations like firms, churches, or armies would not be possible if power relations among people were like the pecking orders of chickens. If transitive relations are also antisymmetric, like >, they establish hierarchies of high and low objects.

WEAK ORDERS—POWER AND HIERARCHY

Definition. A *weak order* relation is antisymmetric and transitive.

Weak orders are very important because they describe power hierarchies in human organizations. If person a is a superordinate of person b and b is a superordinate of c, then a is a superordinate of c. This and the relation of being a descendent of another are the best examples of weak orders among human relations.

This property of transitivity is not true for lots of other human and animal antisymmetric relations. For example, pecking, a kind of dominance relation among chickens, is antisymmetric but not transitive; a can peck b, b can peck c, but c can peck a. The results in a basketball tournament or league need not be transitive. It can happen that a beats b, b beats c, and *c* beats *a*. Informal power and influence are not transitive. Person *a* can have influence over person *b*, person *b* over person *c*, without *a* having any influence whatsoever over person *c*. Your mother influences you, you influence your boyfriend, but your mother probably has no influence over your boyfriend.

EQUIVALENCE RELATIONS

Finally, a relationship that is reflexive, symmetric, and transitive is an *equivalence* relation. Equality, $=$, is an equivalence relation. $a = a$ for every number (reflexivity). If $a = b$, then $b = a$ (symmetry). If $a = b$ and $b = c$, then $a = c$ (transitivity). There are other mathematical equivalence relations.

1. *a* and *b* are straight lines. aRb if *a* is parallel to *b*.
2. *a* and *b* are triangles. aRb if *a* and *b* are similar. Similar triangles may not be the same size, but they must have the same angles.
3. *a* and *b* are triangles. aRb if *a* and *b* are congruent. *a* and *b* are congruent if they coincide exactly when superimposed. They are the same except in different locations.

Now consider some mathematical relations that are not equivalence relations.

1. *a* and *b* are straight lines (of infinite length). aRb if *a* and *b* intersect. R is reflexive and symmetric, but it is not transitive. For example, consider a line *b* that intersects two parallel lines *a* and *c*.
2. *a* and *b* are points on a straight line. aRb if *a* and *b* are less than one inch apart. This relationship is not transitive.

With respect to relations among people, having the same gender or being of the same race or voting for the same party in a presidential elections are all equivalence relations. In general, if aRb means that *a* and *b* both belong to one of a set of mutually exclusive categories, categories that do not overlap, then R is an equivalence relation. If aRb means that *a* and *b* are in the same categories, then it is obvious that aRa, aRb implies bRa, and aRb and bRc imply aRc. The categories that correspond to the equivalence relation are called equivalence classes.

Every equivalence relation corresponds to a set of equivalence classes, even if the equivalence classes are not used to define the equivalence relation. Later in this chapter we will be discussing equivalence relations that are not defined in terms of membership in the same category. We will prove these by showing that if two equivalence classes overlap at all, they must be equal to each other.

Proposition. Let R be an equivalence relation and let $R(a)$ be the set of all objects equivalent to object a; $x \in R(a)$ means xRa. Then, for any a and b, either $R(a)$ and $R(b)$ are disjoint or they are equal.

Proof. Let c be contained in $R(a) \cap R(b)$. Then cRa and cRb. By the symmetry of R, aRc. By the transitivity of R, aRc and cRb imply aRb. Let d be in $R(a)$. dRa and aRb implies, by transitivity, that dRb, so d is in $R(b)$. Therefore $R(a)$ is contained in $R(b)$. Similarly, $R(b)$ is contained in $R(a)$, and so $R(a) = R(b)$. □

Notice that if the categories are not mutually exclusive, then the relation "belonging to the same category" is not an equivalence relation. For example, consider the following relation: aRb if a and b have the same occupation. Some people have two occupations (e.g., taxi driver and actor). Suppose that a is exclusively a taxi driver, c is exclusively an actor, and b is both. Then aRb and bRc but not aRc.

Recently the census has faced a new and difficult issue. The last census had four racial categories: black, white, Asian and Pacific Islander, and American Indian and Alaskan Native. These are treated as mutually exclusive categories; in other words, "is of the same race" has been treated as an equivalence relation. But, with the growth in the number of intermarriages, more and more people feel that they have two racial identities. If person c has mother m and father f of different races and R is the relation "is of the same race as," then for these people mRc and fRc but fRm is not true. Transitivity is violated and racial classes are no longer mutually exclusive equivalence categories. The Bureau of the Census is now pondering what do about the next census.

STRUCTURAL EQUIVALENCE

Sociologists have long been interested in the "roles" people play in groups. A *role* describes a person's behavior and position in a group. For example, a "leader" of a group typically makes decisions for the group as a whole (her behavior) and is respected by other group members (her position). Two people in a group occupy the same role if their behavior and positions are the same. In this section we will describe a mathematical formulation of the concept "position."

Consider the relation R, "respects," among five people, which is given in Figure 4.5.

Notice that relation R is not symmetric, reflexive, or transitive, but we are going to define another relation that has all these properties. Consider the following definition: for any positions x and y, xCy if x chooses (respects) exactly the same set of people as y and is chosen by the same set as y. It should be clear that this relation is symmetric (if x chooses the same set as y, then y chooses the same set as x and if x is chosen by the same

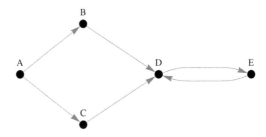

Figure 4.5. A relation that is not symmetric, transitive, or reflexive.

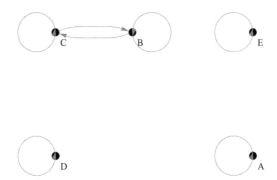

Figure 4.6. Structurally equivalent positions in Figure 4.5.

set as y then y is chosen by the same set as x), reflexive (every x chooses and is chosen by the same set as himself), and transitive. Therefore, C is an equivalence relation generating equivalence classes. The equivalence classes for C are {a}, {b,c}, {d}, and {e}. The graph for the relation C is given in Figure 4.6.

The small circles represent the relation's reflexivity. Note that C is reflexive, symmetric, and transitive, while R has none of these properties.

Definition. xSy (position x is *structurally equivalent* to position y) with respect to relation R if the following two conditions hold: xRw if and only if yRw; zRx if and only if zRy.

Two positions are structurally equivalent if they choose exactly the same set of people *and* are chosen by exactly the same set of people.

Structurally equivalent sets of individuals occupy the same niche even if they have no direct relation to each other. Suppose, for example, that two individuals have almost identical automobiles they wish to sell. They attract the same buyers, and there is no commercial relationship possible between them. Instead they may be competitors. Or, they may collude and agree not to compete with one another.

Ronald Burt, a sociologist, has suggested that under some circumstances information in groups spreads by contagion not between individuals who are connected to one another and talk to one another but between individuals who are structurally equivalent to one another. Suppose that individuals are competing with one another to be experts in some topic: sociology professors who specialize in networks. Suppose that one of them hears that another has mastered a new technique for analyzing social networks. Might she be eager to master this technique herself, to maintain her status as an expert? Burt has applied these ideas to data on the diffusion of the use of a new antibiotic among a set of doctors in some small communities. If two doctors talked with one another, they were not particularly likely to adopt at the same time. However, if they consulted with the same set of other doctors, perhaps as rivalrous experts, they were more likely to adopt simultaneously.

Burt (1995) has also found the same idea to be useful in other context. His book *Structural Holes* is a description of Machiavellian strategies to use in organizations, in particular how to pick your contacts strategically. For example, if you have a limited amount of time to devote to others, maintaining connections to two people who communicate with one another is a waste of your time; you're likely to hear the same information from both of them. This redundancy is even greater if these two people are in turn connected to the same other people, so that their sources of information overlap. In other words, Burt suggests that there is little reason for maintaining contact with sets of people who are structurally equivalent; as a rational individual who wants to get ahead, you should try to arrange your connections so that they are all structurally distinct, giving you access to different parts of the network.

TRANSITIVE CLOSURE: THE SPREAD OF RUMORS AND DISEASES

We want to define a new relationship T_R, called the transitive closure of R, as follows:

Definition. aT_Rb if and only if aRx_1, x_1Rx_2, x_2Rx_3, ..., x_nRb for some sequence of vertices $x_1, x_2, \ldots, x_n$.

For example, if R is the relationship "is bigger by one" among the whole numbers, T_R is the relation "is greater than." $5T_R2$ because $5R4$, $4R3$, and $3R2$. If R is the relation "is a parent of" among people, T_R is the relation "is an ancestor of." If R is the relation "is direct superior of" in an organization, then T_R is the relation "is a superior of."

One surprising conclusion is that T_R is transitive even if R is not. For example, let R be the relationship "is the parent of." It is clear that this relationship is not ordinarily transitive (Do you see why?), but T_R, "is the

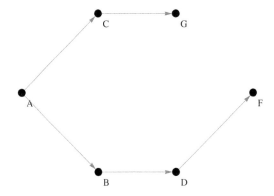

Figure 4.7. An antisymmetric relation.

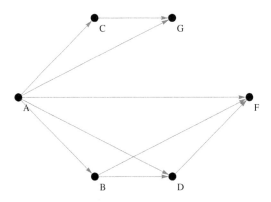

Figure 4.8. Transitive closure of 4.7.

ancestor of," is transitive. On the other hand, if R is already transitive, then $R = T_R$. Figure 4.7 shows a hypothetical authority relation in an organization and Figure 4.8 its transitive closure. Notice that R is not transitive but T_R is.

The transitive closure T_R of a relationship R can connect vertices that are quite distant from one another. I once saw a movie that traced the path of one particular $20 bill. It traced a path all over society, from the height to the depths. These people were all connected by a relation. If R is the relation "gave the bill to or received the bill from," then all these individuals, most of whom did not know each other, were connected by the transitive closure of R.

Now consider the transmission of a rumor, aRb if a tells the rumor to b, aT_Rb if b hears, eventually, the rumor spread by a. Persons a and b can be completely unknown to one another, not even having any friends in common or any friends who are friends with each other, and yet they can be exposed to the same rumor.

Table 4.3.
Chapter glossary

$S \times S$	The Cartesian product of the set S with itself, consisting of all ordered pairs (a, b) where $a \in S$ and $b \in S$.
Relation	A dyadic relation R is a subset of the set $S \times S$.
aRb	$(a, b) \in R$, a relation.
Symmetry	R is a symmetric relation if whenever $(a, b) \in R$, then $(b, a) \in R$ also. In other words, aRb implies bRa.
Antisymmetry	A relation R is antisymmetric if aRb and bRa implies $a = b$.
Reflexivity	A relation R is reflexive if (a, a) is in the relation for all a in S (or, equivalently, aRa for all a in S).
Transitivity	A relation R is transitivie if whenever (a, b) and (b, c) is in the relation, so is (a, c).
Weak order	A weak order relation is antisymmetric and transitive.
Equivalence	A relationship that is reflexive, symmetric, and transitive is an equivalence relation.
Structural equivalence	Two positions x and y are structrually equivalent with respect to a relation R if the following two conditions hold: xRw implies yRw, and vice versa; zRx implies zRy
T_R	The transitive closure of a relation R. aT_Rb if and only if $aRx_1, x_1Rx_2, x_2Rx_3 \ldots, x_nR_b$ for some sequence of vertices $x_1, x_2, \ldots, x_n$.

To take another pertinent illustration, consider the spread of a sexually transmitted disease. Let aRb mean that a has had sex with b. R is a symmetric and intransitive relation. In looking at the spread of sexually transmitted disease, we are interested in a subset of the transitive closure T_R of R.

Chapter Demonstrations

- *Transitivity Game* tests you on your understanding of transitivity.
- *Relations Mini Quiz* tests your understanding of the concepts of reflexivity, symmetry, transitivity, weak orderings, and equivalence relations.

EXERCISES

1. For each of the following relations, indicate whether or not it is necessarily symmetric, reflexive, and transitive. A check in the appropriate box means that the relation always has that property. The set S that persons a and b belong to is the set of all persons.

	Symmetric	Antisymmetric	Reflexive	Transitive
a is the mother of b				
a is the ancestor of b				
a is richer than b				
a and b have attended exactly the same schools				
a likes b				
a is married to b				
a and b have the same best friend				
a is b's first cousin				

2. Use the *Relations Mini Quiz* demonstration until you get three no-error trials in a row.

CHAPTER 5

Networks and Graphs

In the last chapter we borrowed from language used to describe mathematical relations to describe relations among people. For example, human relationships, like authority in organizations, can be transitive in the same sense as the "greater than" relation between numbers. In this chapter we introduce another branch of mathematics that has proven to be very useful in describing relations between people: *graph theory*. Graph theory, contrary to what you might think, has nothing to do with statistical plots of data. Graph theory is all about points, called *vertices* or *nodes* (we will use these terms interchangeably), and the lines connecting them, called *edges* or *arcs*. Graphs are diagrams of these vertices and lines. Graphs are diagrams of networks because the vertices are the objects in the network (people, countries, computers, etc.) and the edges are relationships. For example, consider the graph in Figure 5.1, known as a star.

This is a graph with five vertices and four edges. Vertices connected by an edge are said to be adjacent to one another. The number of vertices adjacent to the given vertex is called its *degree*. In this star graph one vertex has degree four and four vertices have degree one.

We can add some flesh to this example if the vertices are people and the edges are some symmetric relation among these people. For example, the relation might be mutual friendship. Figure 5.2 shows a situation in which one person has four friends who are not friends with each other. The degree here has a simple interpretation: it is a measure of each person's popularity. This group is all about Eve—she's the most popular.

This is not a very tight group of friends; most of the pairs are not friends of each other. There are only four friendships among these five people. A little thought will show that there are six missing relations in this group, those among Annie, Drew, Barbara, and Charlotte. A simple measure of the *cohesion* of a group is the number of relations to the total possible relations—4/10 or .40 in this group. This measure of density varies from .00, when no one is connected to anyone else, to 1.00, when all possible pairs are connected. If there are N people in a group, the number of possible relations is $N(N-1)/2$. In this case, $N = 5$ and $5 \times 4/2 = 10$.

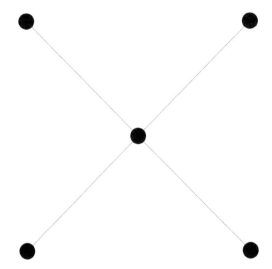

Figure 5.1. A graph with five nodes and four edges.

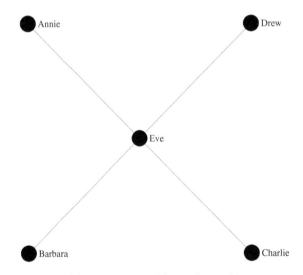

Figure 5.2. A network of five persons and four relationships.

$$\text{Number of possible edges in a graph with N vertices} = \frac{N(N-1)}{2} \quad (5.1)$$

$$\text{Density} = \frac{\text{Number of Edges}}{\frac{N(N-1)}{2}}$$

$$(5.2)$$

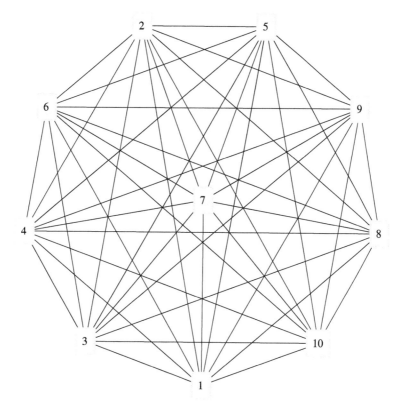

Figure 5.3. A complete network.

We will explore the idea of density further in the chapter on small-world networks.

Notice that although Barbara and Charlotte are not friends, they share a friend. This is important because often in networks things can be transmitted. For example, if Charlotte hears a rumor and friends share information, then Eve will learn of the rumor also. Charlotte may even tell Eve secrets she has learned she has promised to reveal to no one else. If Charlotte catches a cold, Barbara may be exposed to it also through Eve indirectly. A *walk* is a sequence of connected edges that indirectly connects two nodes. Barbara and Charlotte are connected by many different walks: Barbara-Eve-Charlotte, Barbara-Eve-Drew-Eve-Charlotte, Barbara-Eve-Annie-Eve-Drew-Eve-Charlotte, and so forth. Some walks traverse the same edges or the same nodes more than once. Walks that do not repeat either edges or nodes are called *paths*. There is just one path connecting Barbara and Charlotte, although in other more complex networks there may me more than one path connecting two nodes.

A star is one particularly simple kind of network. Another simple type is a complete network, in which all pairs are connected. Figure 5.3 gives a

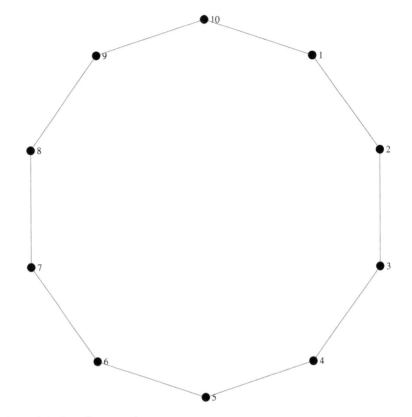

Figure 5.4. A cycle network.

complete network with 10 nodes. Note that even though Node 7 lies in the middle of the graph, the positioning is arbitrary. In a complete network, all nodes are essentially the same in the respect that they have a connection to every other node.

Another simple kind of network is a cycle, a network in which the nodes form a line with the first and last nodes connected to each other. Figure 5.4 gives a cycle network with 10 nodes.

Notice that in this network there are two paths connecting every pair of nodes, but for some pairs the two paths are of unequal length. For example, Nodes 1 and 3 are connected by a path with two edges, 1-2-3, and a path with eight edges, 1-10-9-8-7-6-5-4-3. The length of the shortest path between two nodes is the distance between the nodes, and the shortest paths connecting two nodes are called *geodesics*. Note that some pairs of vertices are connected by two different geodesics.

A grid is a network that can be arranged into a two-dimensional square or rectangle block-like structure. Figure 5.5 gives a grid network with 25 nodes.

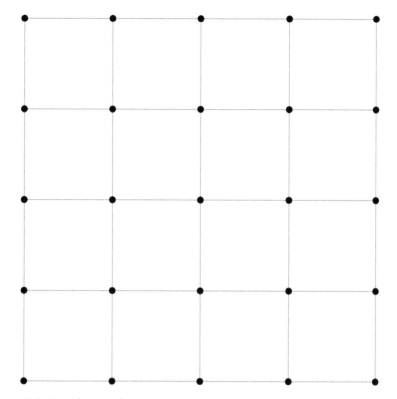

Figure 5.5. A grid network.

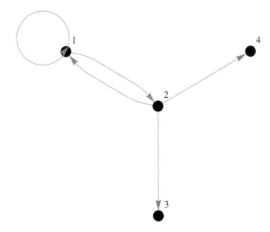

Figure 5.6. A graph that is not simple.

All the networks we will be looking at in this book are *simple*. The word "simple" has a technical meaning. There are no loops (edges connecting a node to itself), and there is at most one edge connecting any pair of vertices. Examine the network in Figure 5.6. There are two reasons why it is not a simple network.

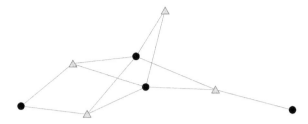

Figure 5.7. A bimodal (bipartite) network.

Some graphs involve two types of complementary nodes in which all connections are between nodes of different types, with none being between nodes of the same type. These are called *bipartite* graphs. For example, many sociologists study networks of directorships of corporations, where there are connections between directors and the corporations they direct. Figure 5.7 shows a hypothetical bipartite dating network between young men and women.

Finding a bipartite network can be an important finding. Suppose you were a newspaper reporter doing a study of expensive Washington restaurants and you observed who ate with whom. You find that almost every pair involved a congressman and a lobbyist. The lobbyist buys an expensive dinner for the congressman in order to influence her vote. You might find that two lobbyists or two congressmen never dine together in these costly restaurants. What's the point! This is the kind of finding you would certainly include in our newspaper article.

It may happen that there is no path between some pairs of vertices; you can't reach one from the other even through indirect paths. This is important because it means that nothing can be transmitted from one vertex to the other. It means, for example, that a contagious disease cannot be transmitted from one indirectly to the other or that a rumor started by one will not reach the other. Sets of vertices that are mutually reachable form a *component* of a graph.

The networks (graphs) we've been looking at thus far have edges, which are bi-directional: A are B are friends of one another, or talk to one another. But some relations can be one directional, and they are represented by digraphs with arcs, not edges. Suppose, for example, that everyone talks about their personal problems with Eve, but she does not reveal any of hers to them. Then the network could be described by the diagram in Figure 5.8, which has four arcs and no edges.

Authority relations in an organizational can also be unsymmetrical. In Figure 5.9 there are three levels to a hierarchy. Positions 1 and 3 both have subordinates. Figures 5.8 is also an example of a trees: it is a network without any cycles. We will talk more about trees in later chapters.

Note that in digraphs the connections of any node are of two types: those that flow in and those that flow out. In Figure 5.8 Eve has four people

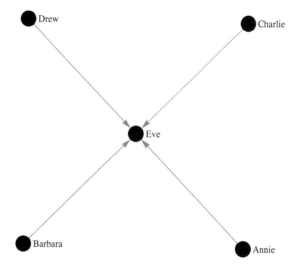

Figure 5.8. A digraph.

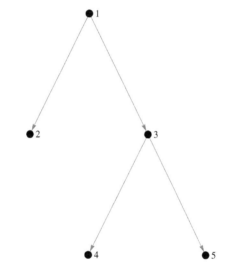

Figure 5.9. A digraph that is a tree.

who talk to her and no one she confides in. In Figure 5.9 Position 3 has one superior and two subordinates. In digraphs we distinguish between in-degree and out-degree. Eve has an *in-degree* of four and an *out-degree* of zero.

EXERCISES

1. Draw all the distinct tree graphs with five vertices. Do they have different densities? Do they have different distributions of degrees? Do some of them have more inequality of degree than others?

Table 5.1.
Graph with 50 nodes.

Average degree	Mostly one component? (yes/no)
0.50	
0.75	
1.00	
1.25	
1.50	
1.75	
2.00	

Table 5.2.
Graph with 100 nodes.

Average degree	Mostly one component? (yes/no)
0.50	
0.75	
1.00	
1.25	
1.50	
1.75	
2.00	

2. Draw a network with five nodes in which the distances between all pairs of nodes are equal.
3. In what size network is the maximum number of edges?
4. What is the maximum number of arcs in a digraph with five vertices? In a digraph with N vertices?
5. What is the maximum distance between two nodes in a network with 10 nodes?
6. Why is the network in Figure 6 not "simple"?
7. Show that the possible components of a graph are mutually exclusive, that they never overlap, and that a vertex cannot belong to two different components.
8. The demonstration Random Graphs creates random networks in which you can control the number of nodes and the average degree. The "random choice" slider will enable you to generate multiple random networks. Experiment with graphs with 50 nodes and an average degree of .50. Is there typically just one component or more than one? Gradually increase the average degree. At what point do most off the random graphs consist of just one component? Now do the same thing for graphs of size 100. Use Tables 5.1 and 5.2.

CHAPTER 6
Weak Ties

BRIDGES

The San Francisco Bay is crossed by seven bridges. The ones first built, the Dumbarton and the San Mateo, are not the most important but rather were built in the early part of the 20th century where the bay is shallower and where the engineering challenges were not as great. The most important bridge in terms of traffic is the San Francisco–Oakland Bay Bridge, constructed a decade after the first two. Its importance lies not only in the fact that it connected the two largest cities in the Bay Area at the time, Oakland and San Francisco. It is important also because distances between population centers were drastically reduced. San Franciscans not taking a ferry boat could get to Oakland only through a lengthy detour down the peninsula to the San Mateo Bridge and the East Bay from Modesto to Oakland. The trip to Marin County across the Golden Gate north of San Francisco was much worse. The importance of a bridge must be a function of the way it reduces distances in a region.

The concept of a *bridge* is also useful in networks and graphs. Consider the communication network in Figure 6.2.

The deletion of the edge between Vertices 1 and 2 would affect only the distance between themselves; it would increase from 1 to 2, but no other pair of vertices would be affected. The connection between 1 and 3, however, is more important. If it were deleted the distances between 1 and 3, 1 and 4, 1 and 5, and 1 and 6 would be increased. Any information originating at Vertex 1 would take longer to diffuse to the group, and vice versa.

The edge between Vertices 3 and 4 is clearly the most important in reducing distances between vertices; it is the Oakland Bay Bridge of this network. The distances between nine pairs are reduced by its presence, more than any other edge, and the degree of reduction is also the greatest; what would be two disconnected components without it becomes one.

Definition. A bridge is an edge whose deletion from a graph causes the number of components to increase.

Clearly bridges are an extreme example of the effect of one edge. We can order edges in terms of the degree to which their removal or addition

Figure 6.1. The seven San Francisco Bay bridges.

Figure 6.2. A network with a bridge.

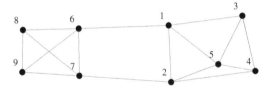

Figure 6.3. A network with local bridges.

affects distances in a graph. Following Granovetter (1973) we will define a *local bridge of degree n* as one whose deletion increases the distance between the two involved vertices from 1 to n. For example, look at Figure 6.3 (from Figure 1 in Granovetter, 1973)

The 6-1 tie is not a bridge; its removal does not increase the number of components. But it is a local bridge of degree 3 because the next shortest path between 6 and 1 is of length 3: from 6 to 7 to 2 to 1.

THE STRENGTH OF WEAK TIES

Now that we've gotten some of the mathematical preliminaries out of the way, let's tackle a substantive issue. How do people find out about jobs? Are relatives, close friends, acquaintances, employment agencies, or newspaper advertisements the most useful sources about jobs? Mark

Granovetter, a sociology graduate student, tried to answer this question in his PhD research. He sent questionnaires to about 200 people in the Boston area who had changed jobs or recently been employed and succeeded in interviewing about 100 of them. All of these were what he called "professional, technical, or managerial" workers; they were not blue-collar workers.

Granovetter was interested in how his respondents found out about their current jobs. Economists like to assume that buyers and sellers in the marketplace, including the marketplace for labor, have "perfect information." This means that all workers know about all the job possibilities for which they are qualified. Such a model might be not too unreasonable if most workers found their jobs through advertisements or through employment agencies. However, Granovetter found that more than half found out about the job opening that led to their current employment through personal contacts; they knew someone who knew about the job opening. Specifically, 56% found their current job through personal contacts, only 19% through advertisements or employment agencies, and 19% through direct application to the firm that hired them (the remaining 7% used other methods or did not answer the question).

Thus, not surprisingly, the model used by economists is misleading. The information people possess about job possibilities is affected by their placement within networks. Let us give you a few examples from Granovetter's book:

Edward A., during high school, went to a party given by a girl he knew. There, he met her older sister's boy friend, who was ten years older than himself. Three years later, when he had just gotten out of the service, he ran into him at a local hangout. In conversation, the boy friend mentioned to Mr. A that his company had an opening for a draftsman; Mr. A applied for the job and was hired. (1973, 76)

Norman G's daughter was in nursery school, where she met the daughter of a lawyer who consequently became friendly with him. When Mr. G. quit his job, the lawyer told him of an opening in the accounting area of a firm which was one of his clients. He applied, and was hired. (1973, 78)

Note how indirect this knowledge is. Granovetter found that many of the paths between the job and the new employee were surprisingly indirect; the person who told her about the job was often someone she did not know very well. Edward found out about the job from his sister's old boyfriend, Norman from another parent in his child's nursery school. Granovetter tried to assess the strength of the relationships between the job seeker and the person who informed him of the job using easy-to-answer questions in interviews and questionnaires. The questions he asked were how often his job seekers saw the person who told them about their new jobs and whether the person was a friend, a relative, or a business contact. Using these questions, Granovetter distinguished between "strong" and "weak" ties. Strong ties exist when people see each other frequently over long periods of time. The relationship is close and intimate. Weak ties are the opposite; weakly tied individuals see one another infrequently, and their relationships are casual rather than intimate.

Granovetter had good reasons for expecting that strong ties would be more useful than weak ties. Those with whom one has strong ties have one's interests at heart. They will give the job seeker whatever valuable information they have. They might also be expected to exert whatever influence they have on behalf of the job seeker. Moreover, the greater frequency of contact means that they have more opportunities to advise the job seeker and have better information about his interests and skills.

Despite all these good reasons why job seekers might be expected to get more and better information from those with whom they had strong ties, the opposite was actually the case. Job seekers almost always found out about the new jobs from people they saw occasionally (less than once a week but at least once a year) or rarely (less than once a year). They also did not find out about their new jobs from friends or relatives (only 31% did). In addition, the jobs that were found through weak ties were on the average better than the jobs found through strong ties; they were better paying, and the workers were more satisfied.

Granovetter did not have adequate information to definitely pinpoint why weak ties were more valuable. He did, however, offer an insightful conjecture, one that has proven to be valuable for others studying the diffusion of information in networks. A person's close friends and relatives who are likely to move in his own social circles are likely to know one another. Therefore, the information they provide is likely to be redundant; what he hears from one friend he is likely to have heard from others as well because they talk to one another and they are themselves exposed to the same sources of information.

On the other hand, one's acquaintances are likely to come from different social circles. They are likely to be different from the job seeker herself and different from each other. Therefore, they are exposed to different sources of information. Each of them will tell the job seeker things she has not heard from other acquaintances or friends.

One's good friends, relatives, and others one sees frequently are likely to have relations with each other, but this is not true of one's acquaintances. Figure 6.4 gives the diagram of this situation.

The people in ego's strong-ties network tend to be tied to one another. The people in ego's weak-ties network may not even know one another. One's weak ties all live in different worlds, and therefore each supplies one with unique information unavailable from other sources. Granovetter suggests that there is a forbidden triad when there are strong ties. If A is connected to B by a strong tie and A is connected to C by a strong tie, then it is impossible for there to be no ties between B and C. Hence, the triad

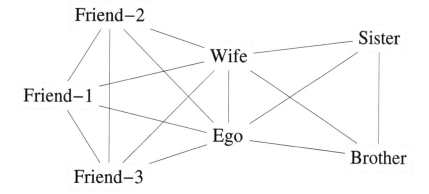

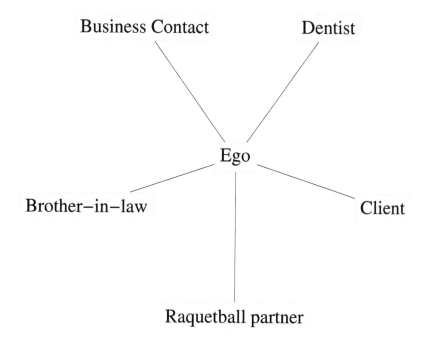

Figure 6.4. Networks of strong and weak ties.

given Figure 6.5 should occur seldom or not at all. We discuss the idea of certain micro configurations not occurring in Chapter 13.

Granovetter is suggesting that strong ties have a weakened form of transitivity; if A has a strong tie with B and B with C, then there will be a ties, strong or weak, between A and C.

The consequence of the forbidden triad is that strong ties will almost never be bridges or local bridges; bridges and local bridges, which provide new information to the individual, will be weak ties exclusively. If strong

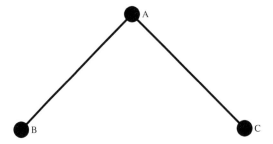

Figure 6.5. The forbidden triad.

ties were a bridge and either one of its nodes had strong ties to a third person, and then the triangle would be complete, the third tie would be strong or weak, and so the initial bridging ties would not in fact be a bridge.

Now, the important point is that information about a job (or about anything) else will diffuse more widely and more quickly because of nontransitive weak ties, which are more likely to bridge otherwise unconnected social worlds (bridges) or at least shorten dramatically the distances between nodes (local bridges).

Chapter Demonstrations

Finding Bridges tests your understanding of bridges and local bridges.

EXERCISES

1. In the notebook *Finding Bridges* the edges in random graphs are colored red if they bridges and green if they are local bridges. Generate random graphs until you are able to predict successfully which edges are bridges and which edges are local bridges. Infer from your experiences what criterion the computer is using to identify local bridges.
2. Use the demonstration *Transitivity Game* (Chapter 4) until you make no mistakes. The demonstration lists the number of intransitive triads and the minimum number of arcs to be added to create a transitive network. Why are these two numbers not always the same?
3. Which of the following six relations are likely to be more transitive and which less. Pick out the three more transitive relations.

   (a) Facebook "friend"
   (b) Kin
   (c) Close personal friend
   (d) Coworker in an office
   (e) Share a taxi ride
   (f) Has sex with

CHAPTER 7

Vectors and Matrices

Matrices and vectors are unfamiliar but essential mathematical tools in the social sciences and in science in general. In this book they will be used extensively in the study of networks and in probability models. They will be introduced in this and the next chapter. Please be patient. Their usefulness will become apparent in ensuing chapters.

The familiar single number is a *scalar*. A column of numbers is a *column vector*, and a row of numbers is a *row vector*. A matrix is a rectangular array of numbers that may have more than one row (a horizontal set of numbers) and column (a vertical set of numbers).

A matrix has both rows (horizontal arrays) and columns (vertical arrays). The number of rows and the number of columns give the order of the matrix. For example, suppose we wanted to create a five by five matrix $\mathbf{A}$ of the total dollar value of trade among five countries. The five rows represent the five countries as exporters and the five columns the same countries as importers.

$$\mathbf{A} = \begin{pmatrix} 0 & 50 & 70 & 80 & 30 \\ 20 & 0 & 49 & 30 & 50 \\ 69 & 68 & 0 & 60 & 80 \\ 70 & 20 & 50 & 0 & 10 \\ 10 & 60 & 70 & 5 & 0 \end{pmatrix} \tag{7.1}$$

Every element of a matrix has two subscripts, for the element's row (counting from the top) and for its column (counting from the left). The rows and columns are numbered from the upper left corner of the matrix. Matrix $\mathbf{A}$ has five rows and five columns. Suppose we found out that exports from Country 3 to country 2 were 70 instead of 68. Then we would change $a_{3,2}$ from 68 to 70. Notice that matrices are referred to by capital letters but their elements lowercase letters.

After the change the matrix $\mathbf{A}$ would have the following form:

$$\mathbf{A} = \begin{pmatrix} 0 & 50 & 70 & 80 & 30 \\ 20 & 0 & 40 & 30 & 50 \\ 60 & 70 & 0 & 60 & 80 \\ 70 & 20 & 50 & 0 & 10 \\ 10 & 60 & 70 & 5 & 0 \end{pmatrix} \tag{7.2}$$

Notice that there are zeros on the *main diagonal* of the matrix. The main diagonal of a matrix runs from the upper left to the lower right. You might ask yourself why this is true.

Some countries, which import more than they export, have a trade deficit, and others, which export more than they import, have a trade surplus. Clearly, a country has a trade surplus if the sum of its row, the amount it exports to other countries, is greater than the sum of its column, the amount it imports from other countries. There is a way in mathematics of representing the sums of numbers using the capital Greek letter sigma. For example, suppose that we want to calculate the sum of the first row of the first **A**, which gives the amount exported by Country 1. We would first define a range variable j that runs from 1 to 5, for the columns of **A**, then sum across the column for the first row.

$$\sum_{j=1}^{5} a_{1j} = a_{11} + a_{12} + a_{13} + a_{14} + a_{15} = 230 \tag{7.3}$$

This says "Take the sum of first row of the matrix **A** across the columns referenced by j." The first subscript in the summation refers to the row of the matrix. The second subscript refers to the column. Since j is a range variable, the summation is over the range of values defined by j. To get the sum of the second row, we would write,

$$\sum_{j=1}^{5} a_{2j} = a_{21} + a_{22} + a_{23} + a_{24} + a_{25} = 140 \tag{7.4}$$

To get the sum of all the rows, we could create a new variable i that stands for the rows, and then form a row sum for each row i. The result is a vector of numbers. We could also create a vector **e** of exports in this fashion.

$$\mathbf{e}_i = \sum_{j=1}^{5} a_{ij} = a_{i1} + a_{i2} + a_{i3} + a_{i4} + a_{i5} \tag{7.5}$$

$$\mathbf{e} = \begin{pmatrix} 230 \\ 140 \\ 268 \\ 150 \\ 145 \end{pmatrix} \tag{7.6}$$

$\mathbf{e_i}$ gives the exports for country i. For example, $\mathbf{e}_3 = 268$. The same thing could be done with respect to imports. We could create a vector **m** of imports by summing down each column of **A**.

$$m_j = \sum_{i=1}^{5} a_{ij} \tag{7.7}$$

$$\mathbf{m} = \begin{pmatrix} 160 \\ 198 \\ 230 \\ 175 \\ 170 \end{pmatrix} \tag{7.8}$$

From these two vectors, we can calculate whether each country has a trade surplus or deficit. The surplus $\mathbf{s}$, positive or negative, is the difference between elements of the two vectors.

$$\mathbf{s} = \mathbf{e} - \mathbf{m} = \begin{pmatrix} 70 \\ -58 \\ 38 \\ -25 \\ -25 \end{pmatrix} \tag{7.9}$$

SOCIOMETRIC MATRICES

Now let us look at another example, one in which a matrix tells us who influences whom among a set of 6 friends. In this matrix $\mathbf{P}$, a one means that the row person influences the column person, and a zero means that he does not. $\mathbf{P}$ is an instance of an adjacency matrix. An adjacency matrix is a square matrix of ones and zeros. A one corresponds to the presence of a relationship, and a zero corresponds to its absence. For example, in the matrix $\mathbf{P}$, $p_{ij} = 1$ means that iPj, or i influences j.

$$\mathbf{P} = \begin{pmatrix} 0 & 1 & 1 & 1 & 1 & 1 \\ 0 & 0 & 1 & 0 & 0 & 0 \\ 0 & 1 & 0 & 0 & 0 & 1 \\ 0 & 0 & 0 & 0 & 1 & 1 \\ 0 & 0 & 0 & 1 & 0 & 1 \\ 0 & 0 & 1 & 1 & 1 & 0 \end{pmatrix} \tag{7.10}$$

If we sum the rows of this matrix, we can find out how many people are influenced by each person.

$$\text{Influence} = \begin{pmatrix} 5 \\ 1 \\ 2 \\ 2 \\ 2 \\ 3 \end{pmatrix} \tag{7.11}$$

On the other hand, the sum of the columns gives us how many others each person was influenced by

$$\text{Influenced by} = \begin{pmatrix} 0 \\ 2 \\ 3 \\ 3 \\ 3 \\ 4 \end{pmatrix}. \tag{7.12}$$

Notice how different the row and column sums are. The sum of each row is a possible measure of each person's *power*, since it tells how many people she influences. The sum of each column, however, tells us how many people each person lets influence her. It would be a measure of how subject someone is to influence from others. A person high on power but low on being influenced by others would be a trendsetter. A person high on power but subject to influence from others would be a transmitter of trends.

Now let us look at another example. The following matrices show patterns of friendship and helping among 14 workers in a section of a factory. The **H** matrix shows who helped whom, and the **P** matrix shows which pairs of workers were friends.

$$
\mathbf{P} =
\begin{pmatrix}
0 & 0 & 0 & 0 & 1 & 0 & 0 & 0 & 0 & 0 & 0 & 0 & 0 & 0 \\
0 & 0 & 0 & 0 & 0 & 0 & 0 & 0 & 0 & 0 & 0 & 0 & 0 & 0 \\
0 & 0 & 0 & 0 & 1 & 1 & 0 & 0 & 0 & 0 & 0 & 1 & 0 & 0 \\
0 & 0 & 0 & 0 & 0 & 0 & 0 & 0 & 0 & 0 & 0 & 0 & 0 & 0 \\
1 & 0 & 1 & 0 & 0 & 1 & 0 & 0 & 0 & 0 & 0 & 1 & 0 & 0 \\
0 & 0 & 1 & 0 & 1 & 0 & 0 & 0 & 0 & 0 & 0 & 1 & 0 & 0 \\
0 & 0 & 0 & 0 & 0 & 0 & 0 & 0 & 0 & 0 & 0 & 0 & 0 & 0 \\
0 & 0 & 0 & 0 & 0 & 0 & 0 & 0 & 0 & 0 & 0 & 0 & 0 & 0 \\
0 & 0 & 0 & 0 & 0 & 0 & 0 & 0 & 0 & 1 & 1 & 1 & 0 & 0 \\
0 & 0 & 0 & 0 & 0 & 0 & 0 & 0 & 1 & 0 & 1 & 0 & 0 & 1 \\
0 & 0 & 0 & 0 & 0 & 0 & 0 & 0 & 1 & 1 & 0 & 0 & 0 & 1 \\
0 & 0 & 1 & 0 & 1 & 1 & 0 & 0 & 1 & 0 & 0 & 0 & 0 & 0 \\
0 & 0 & 0 & 0 & 0 & 0 & 0 & 0 & 0 & 0 & 0 & 0 & 0 & 0 \\
0 & 0 & 0 & 0 & 0 & 0 & 0 & 0 & 0 & 1 & 1 & 0 & 0 & 0
\end{pmatrix}
\tag{7.13}
$$

$$
\mathbf{H} =
\begin{pmatrix}
0 & 0 & 0 & 0 & 0 & 0 & 0 & 0 & 0 & 0 & 0 & 0 & 0 & 0 \\
0 & 0 & 0 & 0 & 0 & 0 & 0 & 0 & 0 & 0 & 0 & 0 & 0 & 0 \\
0 & 0 & 0 & 0 & 1 & 0 & 0 & 0 & 0 & 0 & 1 & 1 & 0 & 0 \\
0 & 0 & 0 & 0 & 1 & 1 & 0 & 0 & 0 & 0 & 0 & 1 & 0 & 0 \\
0 & 0 & 0 & 1 & 0 & 0 & 0 & 0 & 0 & 0 & 0 & 0 & 0 & 0 \\
0 & 0 & 1 & 0 & 1 & 0 & 0 & 1 & 0 & 0 & 0 & 0 & 0 & 0 \\
0 & 0 & 0 & 0 & 1 & 0 & 0 & 0 & 0 & 0 & 0 & 0 & 0 & 0 \\
0 & 0 & 0 & 0 & 1 & 0 & 0 & 0 & 1 & 1 & 1 & 0 & 0 & 0 \\
0 & 0 & 0 & 0 & 0 & 0 & 0 & 0 & 0 & 0 & 0 & 0 & 0 & 1 \\
0 & 0 & 0 & 0 & 0 & 0 & 0 & 1 & 1 & 0 & 1 & 0 & 0 & 0 \\
0 & 0 & 0 & 0 & 0 & 0 & 0 & 0 & 0 & 0 & 0 & 0 & 0 & 1 \\
0 & 0 & 0 & 0 & 0 & 0 & 0 & 0 & 1 & 0 & 0 & 0 & 0 & 0 \\
0 & 0 & 0 & 0 & 0 & 0 & 0 & 1 & 0 & 0 & 0 & 0 & 0 & 0 \\
0 & 0 & 0 & 0 & 0 & 1 & 0 & 0 & 0 & 1 & 0 & 0 & 0 & 0
\end{pmatrix}
\tag{7.14}
$$

The sums of the rows of **P** give us the number of friends each person had. The sum of the rows of **H** gives the number of others each person helped. The sum of the columns of **H** give the number of others who helped each person.

Notice that the row and column sums of **H** mean quite different things. A person whose row sum for **H** is large helps many others. He is generous and perhaps good at his job, so that others want his help. A person whose column sum is large is helped by others. He may seem helpless.

Remember that a symmetric relation is one in which aRb implies bRa. Therefore, the adjacency matrix **R** for a symmetric relation R is one in which $aRb = bRa$ for every a and b. This type of matrix is called a symmetric matrix. In a symmetric matrix the ith row is equal to the ith column for every i: $r_{ij} = r_{ji}$ for every i and j. You should satisfy yourself that the matrix **P** is symmetric while the matrix **H** is not.

PROBABILITY MATRICES

As we shall see in the later chapter, matrices are useful ways of analyzing probabilities of changes among sets of states. What you learn in these chapters might help you in your next trip to Las Vegas. Suppose, for example, that you are playing the roulette wheel in Las Vegas, making $10 bets on red or black, and that you will quit once you have lost all your money or have $100. There are 11 states you can be in, having $0, $10, up to $100. The casinos have arranged things so that you are more likely to lose (10/19) each $10 bet than to win (9/19; surprise!). An 11 by 11 matrix can be used to show the probabilities of moving from one state to another.

$$
\begin{pmatrix}
1 & 0 & 0 & 0 & 0 & 0 & 0 & 0 & 0 & 0 & 0 \\
\frac{10}{19} & 0 & \frac{9}{19} & 0 & 0 & 0 & 0 & 0 & 0 & 0 & 0 \\
0 & \frac{10}{19} & 0 & \frac{9}{19} & 0 & 0 & 0 & 0 & 0 & 0 & 0 \\
0 & 0 & \frac{10}{19} & 0 & \frac{9}{19} & 0 & 0 & 0 & 0 & 0 & 0 \\
0 & 0 & 0 & \frac{10}{19} & 0 & \frac{9}{19} & 0 & 0 & 0 & 0 & 0 \\
0 & 0 & 0 & 0 & \frac{10}{19} & 0 & \frac{9}{19} & 0 & 0 & 0 & 0 \\
0 & 0 & 0 & 0 & 0 & \frac{10}{19} & 0 & \frac{9}{19} & 0 & 0 & 0 \\
0 & 0 & 0 & 0 & 0 & 0 & \frac{10}{19} & 0 & \frac{9}{19} & 0 & 0 \\
0 & 0 & 0 & 0 & 0 & 0 & 0 & \frac{10}{19} & 0 & \frac{9}{19} & 0 \\
0 & 0 & 0 & 0 & 0 & 0 & 0 & 0 & \frac{10}{19} & 0 & \frac{9}{19} \\
0 & 0 & 0 & 0 & 0 & 0 & 0 & 0 & 0 & 0 & 1
\end{pmatrix}
\qquad (7.15)
$$

All but two of the main diagonal elements of the matrix are zeros because you either win or lose. The main diagonals in the rows for $0 (the first) and $100 (the last) are 1.00 because these are when you have decided to stop gambling.

In a later chapter we will examine the advantages of using matrices to answer questions like the probability of your ending up either broke or with $100 and how this depends on the amount of money you start with.

THE MATRIX, TRANSPOSED

We know that the element in the ith row and jth column of an adjacency matrix $\mathbf{R}$ gives us the relation between person i and person j.

$$r_{ij} = 1 \quad \text{when } iRj \tag{7.16}$$

The transpose of a matrix $\mathbf{R}$, $\mathbf{R}^T$, reverses the rows and columns.

$$(\mathbf{R}^T)_{ij} = 1 \quad \text{whenever } jRi \tag{7.17}$$

$$(\mathbf{R}^T)_{ij} = 1 \quad \text{whenever } r_{ij} = 1 \tag{7.18}$$

For example, the matrix $\mathbf{H}$ shows who (rows) helped whom (the columns), but the transpose of $\mathbf{H}$ would show who (the rows) was helped by whom (the columns).

We have been discussing the transpose of adjacency matrices, but all matrices, whether adjacency matrices or not, have transposes formed by reversing the rows and columns. The general definition of the transpose is as follows:

Definition. In the transpose $\mathbf{A}^T$ of the matrix $\mathbf{A}$, $(\mathbf{A}^T)_{ij} = \mathbf{A}_{ji}$.

For example,

$$\begin{pmatrix} 3 & -1 & 5 \\ 0 & 2 & 4 \\ 5 & 1 & 3 \end{pmatrix}^T = \begin{pmatrix} 3 & 0 & 5 \\ -1 & 2 & 1 \\ 5 & 4 & 3 \end{pmatrix} \tag{7.19}$$

It should be clear that if a relationship R is symmetric, then $\mathbf{R}^T = \mathbf{R}$ for its adjacency matrix. Also, it should be clear that for any matrix $\mathbf{M}$, $(\mathbf{M}^T)^T = \mathbf{M}$.

Finally, the main diagonal of the $\mathbf{A}$ matrix on the first page of this chapter is filled with zeros because a country cannot export to itself.

Chapter Demonstrations

- *Graphs from Matrices*

EXERCISES

1. The demonstration *Graphs from Matrices* shows how zero-one adjacency matrices can represent networks. Why is only the bottom half of the

matrix (below the main diagonal) represented in the top figure, the one you click on and check?

2. Using this demonstration, create a network in which one person is liked by five others who are not friends of each other. Draw the matrix for this network.

3. Again using the demonstration, create a network in which there are five people, all of whom like each other. Draw the matrix for this network.

CHAPTER 8

Adding and Multiplying Matrices

Matrices can be added and multiplied according to their own special rules. If $\mathbf{A}$ and $\mathbf{B}$ are two matrices, then $\mathbf{A} + \mathbf{B}$ is their sum and $\mathbf{A} \times \mathbf{B} = \mathbf{A.B} = \mathbf{AB}$ is their product. For example, consider the following two matrices giving the amount of trade between countries in two different years.

$$\mathbf{T}^{(1)} = \begin{pmatrix} 0 & 59 & 70 & 80 & 30 \\ 21 & 0 & 40 & 30 & 50 \\ 63 & 66 & 0 & 60 & 80 \\ 68 & 24 & 50 & 0 & 10 \\ 17 & 70 & 90 & 15 & 0 \end{pmatrix} \quad \mathbf{T}^{(2)} = \begin{pmatrix} 0 & 50 & 70 & 80 & 30 \\ 20 & 0 & 40 & 30 & 50 \\ 60 & 70 & 0 & 60 & 80 \\ 70 & 20 & 50 & 0 & 10 \\ 10 & 60 & 70 & 5 & 0 \end{pmatrix} \quad (8.1)$$

Addition of Matrices

For addition of matrices, the dimensions of the two must be exactly the same, or else addition is not defined. $\mathbf{T}^{(1)} + \mathbf{T}^{(2)}$ is a new five by five matrix whose elements are the sums of the corresponding elements: $(\mathbf{T}^{(1)} + \mathbf{T}^{(2)})_{ij} = \mathbf{T}^{(1)}_{ij} + \mathbf{T}^{(2)}_{ij}$

$$\mathbf{T}^{(1)} + \mathbf{T}^{(2)} = \begin{pmatrix} 0 & 109 & 140 & 160 & 60 \\ 41 & 0 & 80 & 60 & 100 \\ 123 & 136 & 0 & 120 & 160 \\ 138 & 44 & 100 & 0 & 20 \\ 27 & 130 & 160 & 20 & 0 \end{pmatrix} \quad (8.2)$$

Let us look again at the matrices of liking and helping that we used in the past chapter.

$$\mathbf{P} + \mathbf{H} = \begin{pmatrix} 0 & 0 & 0 & 0 & 1 & 0 & 0 & 0 & 0 & 0 & 0 & 0 & 0 & 0 \\ 0 & 0 & 0 & 0 & 0 & 0 & 0 & 0 & 0 & 0 & 0 & 0 & 0 & 0 \\ 0 & 0 & 0 & 0 & 2 & 1 & 0 & 0 & 0 & 0 & 1 & 2 & 0 & 0 \\ 0 & 0 & 0 & 0 & 1 & 1 & 0 & 0 & 0 & 0 & 0 & 1 & 0 & 0 \\ 1 & 0 & 1 & 0 & 0 & 1 & 0 & 0 & 0 & 0 & 0 & 1 & 0 & 0 \\ 0 & 0 & 2 & 0 & 2 & 0 & 0 & 1 & 0 & 0 & 0 & 1 & 0 & 0 \\ 0 & 0 & 0 & 0 & 1 & 0 & 0 & 0 & 0 & 0 & 0 & 0 & 0 & 0 \\ 0 & 0 & 0 & 0 & 1 & 0 & 0 & 0 & 1 & 1 & 1 & 0 & 0 & 0 \\ 0 & 0 & 0 & 0 & 0 & 0 & 0 & 0 & 0 & 1 & 1 & 1 & 0 & 1 \\ 0 & 0 & 0 & 0 & 0 & 0 & 0 & 1 & 2 & 0 & 2 & 0 & 0 & 1 \\ 0 & 0 & 0 & 0 & 0 & 0 & 0 & 0 & 1 & 1 & 0 & 0 & 0 & 2 \\ 0 & 0 & 1 & 0 & 1 & 1 & 0 & 0 & 2 & 0 & 0 & 0 & 0 & 0 \\ 0 & 0 & 0 & 0 & 0 & 0 & 0 & 1 & 0 & 0 & 0 & 0 & 0 & 0 \\ 0 & 0 & 0 & 0 & 0 & 1 & 0 & 0 & 0 & 2 & 1 & 0 & 0 & 0 \end{pmatrix} \tag{8.3}$$

The matrix $\mathbf{P} + \mathbf{H}$ tells us how many relations there were between pairs of individuals. If a cell is zero, it tells us that that the members of the pair were not friends, nor did the row person help the column person. A matrix cell value of one tells us that either they were friends or the row person helped the column person. A matrix cell value of two tells us both that they were friends and that one helped the other. In other words, $\mathbf{P} + \mathbf{H}$ tells us something about the strength of the bonds between pairs of people.

The addition of matrices has two abstract properties shared with the arithmetic addition of numbers. First, addition of matrices, like the addition of numbers, is *commutative*. This means that for any two matrices, $\mathbf{A} + \mathbf{B} = \mathbf{B} + \mathbf{A}$. Note that not all arithmetic operations are commutative. For example, neither subtraction nor division is commutative; $a - b$ is not in general equal to $b - a$, and a/b is usually not equal to b/a.

The next property, associativity, is more subtle. For numbers, we know that $(a + b) + c = a + (b + c)$. This means that if we first add a and b and add c to the result we get the same number as when we add the sum of b and c to a. The same is true for matrix addition: it is always true that $(\mathbf{A} + \mathbf{B}) + \mathbf{C} = \mathbf{A} + (\mathbf{B} + \mathbf{C})$.

MULTIPLICATION OF MATRICES

Suppose that we have two vectors, p and q, representing the price of a set of goods and the quantity that was sold.

$$\mathbf{p} = \begin{pmatrix} 3.43 & .99 & 17.49 & .34 & 2.23 \end{pmatrix} \tag{8.4}$$

$$\mathbf{q} = \begin{pmatrix} 5 \\ 23 \\ 1 \\ 35 \\ 9 \end{pmatrix} \tag{8.5}$$

It's important that **p** is a row vector and **q** a column vector. **p** is also a one by five matrix and **q** is a five by one matrix. How would we go about calculating the total amount that was spent? We would do this by multiplying the price of each good times the quantity of that good that and then adding up the result.

$$\mathbf{p.q} = 3.43(5) + .99(23 + 17.49(1) + .34(35) + 2.23(9) = 89.38 \qquad (8.6)$$

This is exactly how the product of these two matrices is defined. The product is the sum of the products of the corresponding elements. Now let us make **q** into a five by two matrix in which the two columns represent the amounts purchased by two different shoppers.

$$\mathbf{q} = \begin{pmatrix} 5 & 0 \\ 23 & 35 \\ 1 & 0 \\ 35 & 46 \\ 9 & 5 \end{pmatrix} \qquad (8.7)$$

$$\sum_i \mathbf{p}_i \mathbf{q}_{i1} = 89.38 \qquad (8.8)$$

$$\sum_i \mathbf{p}_i \mathbf{q}_{i2} = 61.44 \qquad (8.9)$$

This is exactly how the product of **p** and **q** is defined.

$$\mathbf{p.q} = \begin{pmatrix} 89.38 & 61.44 \end{pmatrix} \qquad (8.10)$$

Now let us change the problem again. Let **p** be the prices at two different stores. **p** is now a two by five matrix.

$$\mathbf{p} = \begin{pmatrix} 3.43 & .99 & 17.49 & .34 & 2.23 \\ 2.43 & 1.07 & 14.33 & .53 & 1.99 \end{pmatrix} \qquad (8.11)$$

$$\mathbf{p.q} = \begin{pmatrix} 89.38 & 61.44 \\ 87.55 & 71.78 \end{pmatrix} \qquad (8.12)$$

More generally, let **A** be a matrix with m rows and n columns and let **B** be a matrix with n rows and p columns. Note that the requirement for matrix multiplication is that the number of columns of the first matrix equals the number of rows of the second. Then **AB** is a matrix with m rows and p columns formed by multiplying the corresponding elements of each row of **A** and each column of **B** and adding up the products. Let $\mathbf{C} = \mathbf{A} \times \mathbf{B}$. Then the elements in the product are defined in the following way.

$$c_{ij} = a_{i1}b_{1j} + a_{i2}b_{2j} + \cdots + a_{in}b_{nj} \qquad (8.13)$$

Now let us work out an example.

$$\mathbf{A} = \begin{pmatrix} 5 & 3 & 2 & 0 \\ 9 & 4 & 8 & 4 \end{pmatrix} \qquad (8.14)$$

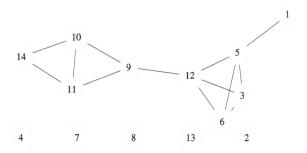

Figure 8.1. Friendships among a set of workers.

$$\mathbf{B} = \begin{pmatrix} 2 & 0 & 1 \\ 9 & 0 & 3 \\ 1 & 4 & 1 \\ 4 & 2 & 7 \end{pmatrix} \tag{8.15}$$

$$\mathbf{A} \times \mathbf{B} = \begin{pmatrix} 39 & 8 & 16 \\ 78 & 40 & 57 \end{pmatrix} \tag{8.16}$$

Surprisingly, matrix multiplication is not commutative. One reason is that sometimes **AB** is defined but **BA** is not. For example, **BA** is not defined for the two matrices above. Even if **AB** and **BA** are both defined, usually, **AB** and **BA** are not equal. We will see examples of this in later chapters. However, matrix multiplication is associative; $(\mathbf{AB})\mathbf{C} = \mathbf{A}(\mathbf{BC})$ is always true if the appropriate multiplications are defined. Use the demonstration *Matrix Multiplication* to familiarize yourself with matrix multiplication. It can be used to generate many examples in which you can strengthen your familiarity with matrix multiplication.

MULTIPLICATION OF ADJACENCY MATRICES

Now let us look at a diagram (Figure 8.1) that corresponds to the **P** matrix of friendship choices in the bank wiring room of Roethlisberger et al. (1939).

The adjacency matrix for this graph is the 14 by 14 matrix **P**. $\mathbf{P}_{ij} = 1$ only if i and j are friends. If P is the relationship as well as the matrix, then $\mathbf{P}_{ij} = 1$ only if iPj. Now let's look at a different relationship, P^2, having friends in common. iPj only when i and j have a common friend; they may or may not be friends with each other. More specifically, iP^2j if and only if there is some person k such that iPk and kPj. Person k is the friend they have in common. Why we call this relationship P^2 will be clearer soon.

Looking at the Figure 8.1 we can see that Nodes 3 and 5 have friends in common (two of them), while 3 and 14 do not. We could graph the relationship $\mathbf{P}^2$. In this graph there would be a line between Nodes 1 and 3 even though they are not themselves friends. Similarly, there would not be a line between 1 and 5 because although they are friends they do not share a friend.

We could also create an adjacency matrix for this relationship. What I want to show is that the adjacency matrix for the relation P^2 is (almost) equal to the matrix $\mathbf{P}^2 = \mathbf{P} \times \mathbf{P}$. Consider again the definition of the relationship $P^2 : iP^2j$ if and only if there exists a person k such that iPk and kPj. Now consider the definition of the i, j element in the matrix $\mathbf{P}^2 = \mathbf{P} \times \mathbf{P}$.

$$P_{ij}^2 = \sum_k P_{ik} p_{kj} \tag{8.17}$$

This will be the sum of the number of instances in which i and k are friends and k and j are friends. Thus, P_{ij}^2 is just the number of common friends k of both i and j. $\mathbf{P}^2$ is not quite the adjacency matrix for the relationship P^2 because it can contain elements greater than one (when two people have more than one friend in common). But it can easily be converted into an adjacency matrix by changing all values that are greater than one into one.

$$\mathbf{P}^2 = \begin{pmatrix} 1 & 0 & 1 & 0 & 0 & 1 & 0 & 0 & 0 & 0 & 0 & 1 & 0 & 0 \\ 0 & 0 & 0 & 0 & 0 & 0 & 0 & 0 & 0 & 0 & 0 & 0 & 0 & 0 \\ 1 & 0 & 3 & 0 & 2 & 2 & 0 & 0 & 1 & 0 & 0 & 2 & 0 & 0 \\ 0 & 0 & 0 & 0 & 0 & 0 & 0 & 0 & 0 & 0 & 0 & 0 & 0 & 0 \\ 0 & 0 & 2 & 0 & 4 & 2 & 0 & 0 & 1 & 0 & 0 & 2 & 0 & 0 \\ 1 & 0 & 2 & 0 & 2 & 3 & 0 & 0 & 1 & 0 & 0 & 2 & 0 & 0 \\ 0 & 0 & 0 & 0 & 0 & 0 & 0 & 0 & 0 & 0 & 0 & 0 & 0 & 0 \\ 0 & 0 & 0 & 0 & 0 & 0 & 0 & 0 & 0 & 0 & 0 & 0 & 0 & 0 \\ 0 & 0 & 1 & 0 & 1 & 1 & 0 & 0 & 3 & 1 & 1 & 0 & 0 & 2 \\ 0 & 0 & 0 & 0 & 0 & 0 & 0 & 0 & 1 & 3 & 2 & 1 & 0 & 1 \\ 0 & 0 & 0 & 0 & 0 & 0 & 0 & 0 & 1 & 2 & 3 & 1 & 0 & 1 \\ 1 & 0 & 2 & 0 & 2 & 2 & 0 & 0 & 0 & 1 & 1 & 4 & 0 & 0 \\ 0 & 0 & 0 & 0 & 0 & 0 & 0 & 0 & 0 & 0 & 0 & 0 & 0 & 0 \\ 0 & 0 & 0 & 0 & 0 & 0 & 0 & 0 & 2 & 1 & 1 & 0 & 0 & 2 \end{pmatrix} \tag{8.18}$$

$P_{5,3}^2 = 2$ because 5 and 3 have two friends in common, 6 and 12. The numbers on the main diagonal have a different interpretation. It's not too difficult to show that they are simply the number of friends each person has.

$$P_{ij}^2 = \sum_k p_{ik} p_{ki}$$

$$= \sum_k p_{ik} p_{ik} \qquad \text{because } \mathbf{P} \text{ is a symmetric matrix}$$

$$= \sum_k p_{ik}^2 = \sum_k p_{ik} \qquad \text{because } \mathbf{P} \text{ is binary} \tag{8.19}$$

Therefore, the main diagonal of the square of the symmetric matrix **M** shows the number of friends of each person.

The interpretation of the elements of R^2 is subtly different when R is not a symmetric relationship. Consider the adjacency matrix **R** for any relation R. R^2_{ij} is the number of walks of length 2 from position i to position j. A walk of length 2 from i to j means that there is at least one person k such that iRk and kRj. But then $R_{ik} = R_{kj} = R_{ik}R_{kj} = 1$ for each such intermediate person k. So, R^2_{ij} is the number of walks of length 2 from i to j.

The most general statement is as follows:

Let **R** *be the adjacency matrix for a relationship* R, *which may or may not be symmetric. Then,* R^n_{ij} *is the number of walks of length* n *from person* i *to person* j.

The Mathematica demonstration *Matrix Powers* gives you a chance to play around with these ideas. A random network is created, its adjacency matrix is shown, and then a graph of walks of various lengths is also shown. These graphs of indirect connections between nodes should make sense to you.

LOCATING CLIQUES

High schools, and probably all human groups, are rife with cliques, groups of friends who form a group and who exclude others. "Cliquishness" means exclusivity; members of a clique are not friends with those outside the clique. Researchers who study networks often want to locate cliques within the network. For example, to what extent are high school cliques based on race, or class, or sex, or participation in sports? Similar questions can be asked about the workplace. Network researchers may have information only about who likes whom, and they may have to surmise or guess where the cliques are based on this limited information. Many definitions have been developed, and in this section we will explore two possible definitions of cliques, one of which is extremely demanding and the other of which is very loose.

Definition. Let R be a symmetric relationship. A *clique S* is a set of people for whom aRb for all a,b $\in$ S and S is not a proper subset of some larger clique. [Note: A set S is a proper subset of a set T if all elements of S are in T and there is at least one element of T that is not in S.]

There are two parts to this definition, and both parts are important. Consider the following hypothetical graph of friendship relations among a group of people.

The three cliques are {1, 2, 3, 4}, {3, 5, 6}, and {6, 7}. {1, 2, 3} is not a clique because it forms a subset of the larger clique {1, 2, 3, 4}. Notice that Person 3 belongs to two different cliques.

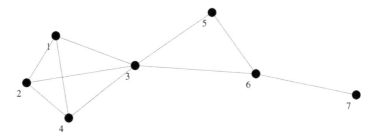

Figure 8.2. A network with two cliques.

A component, on the other hand, is a set of people who are connected by ties of friendship even if they are very indirect. Two people are in the same component if they are friends, or if they share friends, or if their friends are friends, and so forth. In terms of graphs, two people, by this definition, are in the same component if there is a path of any length connecting them.

Definition. Let R be a symmetric relation. A *component* is a set of people all of whom are connected directly or indirectly by paths of any length and who are not a subset of a larger component.

In the above diagram there is one component consisting of all seven individuals. Individuals 1, 2, 3, and 4 do not form a component because they are a subset of a larger component, the set of all seven individuals.

One mathematical consequence of the definition of components is that being in the same component is an equivalence relation. That is to say, it is symmetric, reflexive, and transitive.

Claim. Let aWb mean that there is a path of friendship choices from a to b. If friendship R is symmetric and reflexive, then W (belonging to the same component) is an equivalence relation.

W is reflexive because R is reflexive. Similarly, since R is symmetric, if there is a path from a to b, there also exists a path in the opposite direction from b to a (if R is not symmetric, then W need not be symmetric). Finally, W is transitive. Remember that W is transitive is whenever aWb and bWc then aWc. Suppose there is a path in a graph from a to b and a path from b to c. Must there be a path from a to c? Of course! There must be at least one path from a to c, one that passes through b. There may be others as well.

Any equivalence relation breaks a set up into equivalence classes, separate sets of individuals who are not related to each other. Look at the diagram of friendships in the bank wiring room. There is one large component with nine members and five smaller ones consisting of just one person. These five people are "isolates."

There is a way to detect components using matrix multiplication. Remember that **R** to the power k gives the number of walks of length k connecting every pair of persons. If there are n people in a group, the longest nonredundant path (a path that does not backtrack on itself and does not pass through any person more than once) can be at most of length $n - 1$. This does not mean, however, that a path of length $n - 1$ will exist or that it must be nonredundant even if it does. For example, among the 14 individuals in the bank wiring room, the longest possible nonredundant path would pass through each person once and would be of length 13. In the bank wiring room, such a path does not exist since some workers are isolates and are not connected to anyone else. But, if we added up the first $n - 1$ powers of the matrix, we would get the total number of walks connecting each pair. By summing up the first $n - 1$ powers, we are guaranteed of not missing any walks. If this number is zero, the members of the pair are unconnected and must belong to different component. If it is greater than zero, they must belong to the same component.

Let's look at the seven-person group described by Figure 8.1. We could see by eye that they formed one component. This can also be demonstrated by looking at powers of the matrix of their relationships.

$$P = \begin{pmatrix} 0 & 1 & 1 & 1 & 0 & 0 & 0 \\ 1 & 0 & 1 & 1 & 0 & 0 & 0 \\ 1 & 1 & 0 & 1 & 1 & 1 & 0 \\ 1 & 1 & 1 & 0 & 0 & 0 & 0 \\ 0 & 0 & 1 & 0 & 0 & 1 & 0 \\ 0 & 0 & 1 & 0 & 1 & 0 & 1 \\ 0 & 0 & 0 & 0 & 0 & 1 & 0 \end{pmatrix} \tag{8.20}$$

We need only look at walks up to length 6; if two positions aren't connected by a walk of length 6, they aren't connected at all. So, let's take the sum of the first six powers of the matrix **P**.

$$\sum_{k=1}^{6} P^k = \begin{pmatrix} 336 & 336 & 422 & 336 & 188 & 202 & 58 \\ 336 & 336 & 422 & 336 & 188 & 202 & 58 \\ 422 & 422 & 554 & 422 & 248 & 264 & 86 \\ 336 & 336 & 422 & 336 & 188 & 202 & 58 \\ 188 & 188 & 248 & 188 & 122 & 132 & 44 \\ 202 & 202 & 264 & 202 & 132 & 150 & 46 \\ 58 & 58 & 86 & 58 & 44 & 46 & 20 \end{pmatrix} \tag{8.21}$$

As we can see, there are walks connecting all pairs of points, so they are in one component.

Chapter Demonstrations

- *Matrix Multiplication* demonstrates the rules of matrix multiplication
- *Matrix Powers* shows that a power of an adjacency matrix shows the number of walks of that length connecting pairs of vertices.

EXERCISES

1. Add the following pairs of matrices, if possible. If not possible, explain why.

(a) $\begin{pmatrix} 7 & -4 \\ 2 & 2 \end{pmatrix} + \begin{pmatrix} 2 & -1 \\ 6 & 2 \end{pmatrix}$

(b) $\begin{pmatrix} 4 & 3 \\ -1 & 4 \\ -5 & 7 \end{pmatrix} + \begin{pmatrix} 1 & 2 \\ 6 & -4 \\ 7 & 2 \end{pmatrix}$

(c) $\begin{pmatrix} 4 \\ -1 \\ -5 \end{pmatrix} + \begin{pmatrix} 4 \\ -1 \\ -5 \end{pmatrix}$

(d) $(2 \quad 8) + (6 \quad 1)$

(e) $\begin{pmatrix} 1 & 3 & 4 \\ 5 & 2 & 1 \\ 9 & 7 & 2 \end{pmatrix} + \begin{pmatrix} 1 & -1 \\ 3 & -3 \end{pmatrix}$

(f) $\begin{pmatrix} 5 & -3 & -1 \\ 9 & 5 & 5 \end{pmatrix} + \begin{pmatrix} 1 & 3 & 5 \\ 2 & 4 & 6 \end{pmatrix}$

(g) $\begin{pmatrix} 5 & 1 \\ 6 & 1 \end{pmatrix} + 7$

(h) $\begin{pmatrix} 1 & 3 & 4 \\ 5 & 2 & 1 \\ 9 & 7 & 2 \end{pmatrix} + \begin{pmatrix} 1 & 7 & -1 \\ 5 & 2 & -2 \\ 3 & 8 & -3 \end{pmatrix}$

2. Multiply the following pairs of matrices, if possible. If not possible, explain why.

(a) $\begin{pmatrix} 5 & 2 \\ 4 & 3 \end{pmatrix} \cdot \begin{pmatrix} 1 & 0 \\ 0 & 1 \end{pmatrix}$

(b) $\begin{pmatrix} 5 & 6 \\ 0 & 0 \end{pmatrix} \cdot \begin{pmatrix} 5 & 2 \\ 3 & 4 \end{pmatrix}$

(c) $\begin{pmatrix} 0 & 0 \\ 0 & 0 \end{pmatrix} \cdot \begin{pmatrix} 0 & 0 & 0 \\ 0 & 0 & 0 \end{pmatrix}$

(d) $\begin{pmatrix} 1 & 1 & 8 & 6 \\ 6 & 9 & 2 & 3 \\ 8 & 6 & 4 & -7 \\ 2 & 5 & 1 & -7 \end{pmatrix} \cdot \begin{pmatrix} 1 & 0 & 0 & 0 \\ 0 & 1 & 0 & 0 \\ 0 & 0 & 1 & 0 \\ 0 & 0 & 0 & 1 \end{pmatrix}$

(e) $\begin{pmatrix} 5 & 2 \\ 5 & 3 \\ 2 & 3 \end{pmatrix} \cdot \begin{pmatrix} 1 & 5 \\ 2 & 1 \\ 4 & 5 \end{pmatrix}$

(f) $\begin{pmatrix} 5 \\ 5 \\ 2 \end{pmatrix} \cdot \begin{pmatrix} 1 & 3 & 4 \end{pmatrix}$

(g) $\begin{pmatrix} 1 & 1 & 8 \\ 6 & 9 & 2 \\ 8 & 6 & 4 \end{pmatrix} \cdot \begin{pmatrix} 3 & 4 & 5 \\ 4 & 2 & 3 \\ 5 & 4 & 2 \end{pmatrix}$

3. Find the cliques and components in the following structure.

$$\mathbf{P} = \begin{pmatrix} 0 & 0 & 0 & 1 & 1 & 0 \\ 0 & 0 & 1 & 0 & 0 & 0 \\ 0 & 1 & 0 & 0 & 0 & 1 \\ 1 & 0 & 0 & 0 & 1 & 0 \\ 1 & 0 & 0 & 1 & 0 & 0 \\ 0 & 0 & 1 & 0 & 0 & 0 \end{pmatrix} \tag{8.22}$$

CHAPTER 9

Cliques and Other Groups

Humans have evolved to live in groups. In our hunting and gathering past, we evolved in groups of size 100 to 150. Even now we grow up, work, and play mostly in small groups. All the processes that social psychologists study, like leadership, status, norms, and so forth, occur in groups. Social psychologists have spent much time and energy defining the term "group", but at the heart of the different definitions is the idea that a group is a set of people who identify themselves as a unit and who pursue common goals. A small group is a set of people small enough so that they all know each other.

We evolved over thousands of years of prehistory to live in hunting and gathering groups of size 150 or so. This is the milieu in which we are designed to live, function socially, and accomplish goals. A military company is approximately this size. There are occasions in which the basic unit of human behavior is the network; information and influence flow through networks, diseases are transmitted through networks. But often we study networks in order to be able to understand the behavior of groups, which have levels of organization above linked dyads. Groups but not networks have leaders, goals, norms, boundaries, and so on. Groups function like organisms, and one very important use of network tools is to locate potential groups when we cannot observe them directly but all we have is the record of dyadic relationships. We are often in a position of trying to infer the potential members of a group from network data.

Identifying groups can be important. In a study of school children, one may want to know the degree to which social groups of children are based on class or race or sex and how this changes or varies by school. If one is studying a social movement that occurred in the past, one may want to identify the groups that led the social movement. Someone studying the U.S. economy may want to know if there are groups of firms that are more powerful because they arrive at common economic and political policies coordinated by having common members of their boards of directors.

However, although we may be interested in groups, they may be difficult or impossible to actually observe in action. Groups of businessmen who are planning which candidates to back in an upcoming election do not hold public meetings. Groups plotting a revolution also may try to keep their meetings secret from the authorities. Groups of children will socialize away from school rather than just in the classroom or the playground. Sometimes we have to infer the existence of groups from signs they leave or from data that we've collected for other purposes, and these data are often network data. So, we will explore ways of inferring the likely existence of a group from network data. For example, the following types of network data may be available.

1. Lists of the boards of directors of large corporations
2. Records of email correspondence within an organization
3. Results of a questionnaire in which children are asked who their friends are
4. A government may be able to obtain telephone records or open mail to identify groups of dissidents
5. Records of which scientists coauthored papers
6. Information on which other scientists are cited in publications
7. A questionnaire in which the researcher asks who likes whom

To infer the possible existence of a group from network data, we look at sections of the network in which ties are especially dense and possible borders between groups that links do cross. These dense areas and borders are the traces that groups leave. Once we have located these traces we can move on to see if the groups we've defined actually behave like groups.

The rather weak starting point for where to look for a real group is a component. It's hard to imagine how a set of people can engage in any coordinated activity if there are subsets with no connections between them at all.

The dictionary definition of a clique is an exclusive circle of people with a common purpose. It often connotes snobbery and arrogance, but the term also has an emotionally neutral and specific meaning in graph theory that is useful in studying social networks.

Definition. A *clique* is a maximal set of nodes between which all edges exist.

If a set of individuals all choose each other as friends and they are not a subset of some larger set with the same property, the set forms a clique. If there are 10 individuals who work in an office together and all 45 $(10 \times 9/2)$ pairs see each other sometimes outside of work (a symmetric relation), then they form a clique with respect to this relationship. The density in a clique is 1.00.

You can also play around with a Mathematica demonstration available for this chapter: *Finding Cliques*. The first of the two sliders determines the size and density of the random network that is generated and the second its density. All the cliques are diagrammed by enlarging the circles of their members. Click on the diagram to move from one clique to another, if there is more than one. There may be no cliques, in which case no nodes will be enlarged. When all the cliques have been shown, you will be returned to the first clique. We will discuss the meaning of the controls later.

Of course a set of people may all know each other and yet there may be no group at all. Suppose that Albert is a coworker of Bernice, Bernice plays squash with Charles, and Charles and Albert are active together in a philatelist club. A clique is only a clue that a set of people capable of organized activity is present. On the other hand, a clique is a very stringent criterion that may miss real groups. Suppose, for example, that 10 coworkers regularly play basketball with each other outside of work but that five pairs of them don't like each other. Because not every tie among the 45 possible ties exists, you will miss this group. There are many different ways of loosening the definition of a clique so that it is not quite so demanding, and we will cover two of them here: N-cliques and K-plexes.

Definition. An *N-clique* is a maximal set of nodes all pairs of which are at a distance of no greater than N from each other.

In other words, all the members of a 2-clique need not be friends, but they must at least have a friend in common.

Definition. A *K-plex* would be a clique except for the absence of up to K missing edges.

Using *Finding Cliques* you should familiarize yourself with these relaxations of the concept of a clique. You can decide the allowable maximum distances between those in the same grouping (the N-clique idea). You can also select the maximum number of missing edges in a grouping (the K-plex idea). You can also create criteria that combine the K-plex and N-clique idea: what are the groupings in which all pairs but possibly one are connected by paths of length two or less? You can also determine the minimum size of the groupings you want to look at. Repeatedly clicking on the network diagram shows all the groupings that satisfy the criteria.

BLOCKS

One characteristic with the clique identification techniques we have examined is that they produce overlapping groups. There may be conditions under which this makes sense. There is no reason to think that informal social groupings are disjoint; someone can belong to more than one circle of friends in different areas of her life. On the other hand, there are groupings in which it makes sense that someone can belong to only one: an adolescent might belong to only one gang, a politician to just one political party. In these cases we might want to use network data to divide people into disjoint cohesive sets.

What we want are groups that are as nearly disjoint as possible. They should be internally cohesive while having as few ties between them as possible. We want to maximize the internal density of groups relative to some baseline, which could be the statistically expected number of internal connections if there were no real groups. Suppose we partition a group into a number of separate communities. Let a_i be the proportion of ties that involve subset i. Then the proportion of ties that are from members of subset i to members of subset i should be a_i^2. Let e_{ij} be the proportion of ties actually between subsets i and j. Then we want a partition that maximizes the following quantity:

$$Q = \sum_i (e_{ii} - a_i^2) \tag{9.1}$$

This is called the community structure partition. It is devilishly difficult to find in large networks in which every partition cannot be examined even with fast computers. There are algorithms, and one of them is implemented in the demonstration *Community Structure*. Play around with it to familiarize yourself with its properties. It "finds" the communities in either random graphs (if "cliques" equals zero) or especially constructed graphs with two underlying communities (if "cliques" equals two).

Chapter Demonstrations

- *Finding Cliques* detects cliques and variants of cliques in networks
- *Community Structure* finds community structure partitions in networks

EXERCISES

1. Show that an N-clique when $N = 1$ must be a clique.
2. Using *Finding Cliques*, what do you observe to be the effect on the size of N-cliques as N increases?
3. Using *Finding Cliques*, what do you observe to be the effect on the size of K-plexes as K increases?
4. Consider the following adjacency matrix.

$$\begin{pmatrix} 0 & 1 & 1 & 0 & 0 & 0 \\ 1 & 0 & 1 & 0 & 0 & 0 \\ 1 & 1 & 0 & 1 & 0 & 0 \\ 0 & 0 & 1 & 0 & 1 & 1 \\ 0 & 0 & 0 & 1 & 0 & 0 \\ 0 & 0 & 0 & 1 & 0 & 0 \end{pmatrix} \tag{9.2}$$

(a) What are the cliques?
(b) What are the 2-cliques
(c) What are the 1-plexes?
(d) What are the 2-plexes?
(e) Without calculation, and using the information that the community structure partition maximizes the density of ties within communities, what would you guess the partition (into two communities) would be for this graph?

CHAPTER 10

Centrality

In an academic department the regular faculty may be away frequently doing their own research or working at home. Because they may be in the department only infrequently to teach and because their research interests may be quite different, they may not see much of each other. On the other hand, the staff of a department, especially the staff in the front office, is there all day five days a week, and they may be the only ones to talk to all members of the department frequently. As a result, they may be the best informed people in the department, better aware than most members of the department about successes, significant personal events, illness, and tragedies.

This was brought home to the senior author (Bonacich) a few years ago when his father died. He had to spend some time in San Francisco settling his father's affairs. Some of his responsibilities had to be transferred to others, and this required coordination with the staff. As a result, the staff all knew about the father's death before most of the faculty. Soon afterward he received a condolence card signed by all the staff, but sympathy from other faculty members trickled in as they found out.

Information flows through a social network, and some individuals will be better positioned within this network to learn more information and to learn it earlier. It should also be clear that in such a network it's not just how many people you talk to but who you talk to; regularly having lunch with someone who is himself well informed will provide you with better and more current office gossip. Being well placed in such a network need not correspond to formal status or rank in the organization. There are firms that specialize in diagnosing communication problems in organizations. An important executive who should be well informed in order to make good decisions may have been shut out of the informal communication network of an organization. There may be no communication between members of two divisions of a firm that should be coordinating their activities.

Information is not the only thing that flows through networks. Electricity flows through power grids, and those who plan such grids should have a good idea of where the system is in danger of becoming overloaded and where it is especially vulnerable to catastrophic failure if a piece of equipment fails. Contagious epidemics also flow through social networks. If a vaccination exists, it may be especially important to vaccinate those whose removal from the network would especially disrupt the spread of the disease, not just those who are most vulnerable. Health workers come in contact with many sick individuals and thus are particularly likely to catch the disease. If children are especially vulnerable, it may be important to inoculate teachers and parents of small children even if they themselves are not especially at risk.

Social scientists and non–social scientists have recently become interested in *social capital*, which refers to the resources available to someone through others in their social networks. Social capital means different things in different circumstances. To a poor person social capital may refer to the availability among her friends of a car she can borrow, free babysitting, money she can borrow in an emergency, and emotional support in times of stress. To an entrepreneur his social capital may refer to the skills available among people he plans to work with.

Rankings of all sorts that we are exposed to daily can be better understood if they are described and understood as networks. For example, consider network search engines. The early search engines were not deficient in the number of webpages they covered. They all had webcrawlers that explored the web by following links from the websites they had discovered. The primary deficiency was in presenting the results. The output from early search engines was page after page of mostly irrelevant websites sprinkled occasionally with a useful site. Who today remembers AltaVista, Magellan, HotBot, or Excite? (You do? Obviously you aren't a young college student.)

Google became the dominant search engine not because it had a better web crawler but because of an algorithm, PageRank, that ordered results in a useful way. The primary innovation behind PageRank is to consider webpages as nodes in a network with links between pages as arcs in a digraph. PageRank locates highly central nodes in this network. These highly central webpages will appear at the top of the list in a Google search. The central idea is to weight links by the importance of the sites that link to them.

Computerized sports ranking systems also make use a network approach. The vertices are sports teams and the arcs represent who beat whom. Victories and losses are weighted according to the rankings of a team's opponent, and the results of one game can have implications that spread throughout the network indirectly affecting the rankings of many other teams.

The centrality measures we examine in this chapter will fall into two broad categories: those that measure how important a node is in the flow through the network of information of some other resource and those that measure the status or prominence of a position by its integration with other prominent positions. The first set of measures are usually based

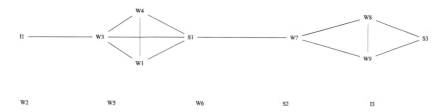

Figure 10.1. Friends in the bank wiring room.

on geodesics, and the latter on connections with other well-connected nodes. A lowly assistant professor with a joint appointment can be the only connection between two departments without being in the ruling clique in either department.

In this chapter we will be using two real networks as examples. The first one involves friendship patterns among a set of workers in a room in the factory, the "bank wiring room" (Figure 10.1). These workers produced telephone equipment. There were wiremen who wired connections, soldermen who soldered the connections, and inspectors (see Homans 1958; Roethlisberger et al. 1939). Lots of things flow through friendship networks: information, reciprocal gifts and favors, and emotional support.

The second network (Figure 10.2), also well studied by social scientists (Padgett and Ansell, 1993), is of marriage ties between important families in the Italian Renaissance city of Florence in the 15th century. Through these ties might flow political, financial, and social support. Social scientists have attempted to account for the power and influence of the Medici family in this period through their positions in this and other networks.

Centrality refers to characteristics of a node's position in a network in which some type of resource flows from node to node. There are many ways in which this rather vague idea could be implemented. We will describe a few of them in this chapter. In what follows we will assume that the network is represented by a symmetric binary (0 and 1) adjacency matrix A. This means that the relation is symmetric and that it either exists or does not exist between any two nodes. There are more complex ways of handling more realistically complex situations (for more methods, see Nooy et al. 2005; Wasserman and Faust 1994), but these assumptions make the presentation manageably simple.

All the measures we are looking at here (with the possible exception of degree centrality) make sense only for sets of vertices that are in the same component, so that their positions within the component can be meaningfully compared. The measures should be computed separately for each component.

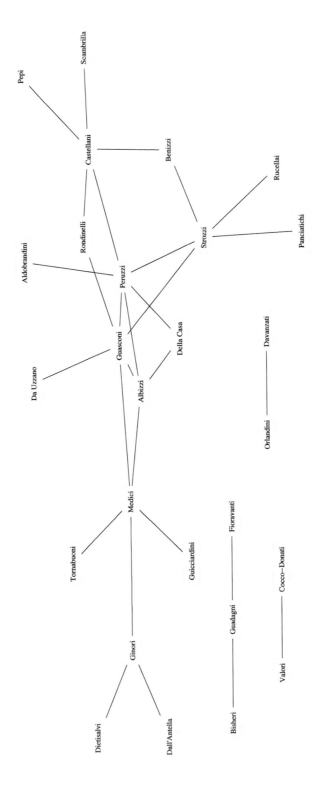

Figure 10.2. Marriage network in Renaissance Florence.

DEGREE CENTRALITY

One of the most basic forms of centrality is degree centrality. Degree centrality, D, is simply a node's total number of connections.

$$D_i = \sum_k a_{ik} \tag{10.1}$$

Everything else being equal (which it probably isn't), the amount of information an individual is exposed to or the resources he has access to is proportional to the number of his contacts. Among the Florentine families, for example, the Guasconi and Peruzzi families were tied to the most other families (six) by marriage, and this gave them an advantage in terms of information and alliances. Degree centrality is often used as a rough measure of popularity or importance. In a network of friends, people with a high degree have the most friends. In a food-chain network where the edges represent predator and prey relationship, species with high degree centrality are important to the overall network.

Degree centrality is popular because it's easy to measure. Large representative random surveys that may not be designed to facilitate network data can include a question or two, such as "How many friends do you have," to approximate degree centrality. However, this aspect of the measure is also its biggest weakness. Degree centrality really isn't a "true" network centrality measure because it does not take into account the structure of the entire network.

GRAPH CENTER

Suppose you ran a large retail store that sold many different items. Your store just received a new barcode scanner that would make life easier for customers who wanted to check the prices on items. Here's the catch— your store received only one of these blasted scanners. Now you are tasked with finding out the best location for the scanner. Well, if you imagine the aisles in your store as a network, you would want to place the scanner at a location that is central to all products. If your store's building is perfectly square, and your aisles follow a neat grid pattern, the most logical place to meet this criterion would be right in the center of the store. In networks, closeness centrality measures this aspect of power. Individuals with high closeness centrality can be thought of having high influence. If they had to pass along information, orders, or resources through their network, they would be able to reach a large number of people fairly quickly.

A crude centrality measure that tries to operationalize the concept above is the graph center.

If everyone in a network has some information to contribute and transmits her information to her neighbors, then the person in the network who possesses the entire group's information first will be the person whose maximum distance from others in the network is least. For example, let's look at the largest component in Figure 10.1, containing nine workers. Just to keep it simple, suppose that each person transmits all he knows to his neighbors in one time period. It will take five time periods for Inspector 1 and Solderman 3 to hear the information known by those on the opposite side of the network. At the other extreme, Solderman 1 and Wireman 7 are no more than three links away from everyone else in the component. They are the centers of this graph. If d_{ij} is the graph distance between nodes i and j, then the center (or centers) is the node k whose maximum distance from other nodes is the minimum. You should be able to show that the Guasconi and Albizzi families are the centers of the network of Florentine families.

$$\text{Vertex } k \text{ is a graph center if } \max_j \{d_{kj}\} = \min_i \max_j \{d_{ij}\} \qquad (10.2)$$

Another advantage of graph centers is that any information they have will become completely distributed to all members of the group the quickest. Solderman 1 and Wireman 7 in Figure 10.1 are in the nest positions to coordinate the activities of a group. If they suggest some group activity, all members of the largest component hear it first. If Solderman 3 were to say that he planned to go to Bar A after work for a drink and Solderman 1 that he planned to go to Bar B and if Figure 10.1 describes communication links, everyone will hear the second suggestion before the first.

CLOSENESS CENTRALITY

Closeness centrality C can be thought of as a refinement the graph center. For each node, it is the reciprocal of the total distance of all other nodes from that node. In other words, given a vertex, its farness centrality is the distance between that vertex and all other vertices—all $(n-1)$ of them. If a vertex is in a remote location, the sum of all these numbers is going to be very big. If the vertex is in a location close to all else, this number will be small. In order to turn farness to closeness, network researchers take the multiplicative inverse (aka divide by 1) of farness in order produce closeness.

$$C_j = \frac{1}{\sum_k d_{jk}} \qquad (10.3)$$

Table 10.1 shows the distances between all pairs of vertices in the largest component of the bank wiring room.

The total distances are given by the vector (23, 17, 16, 17, 14, 18, 18, 13, 24). The closeness centralities are, therefore, (1/23, 1/17, 1/16, 1/17,

Table 10.1.

Distances between vertices in the bank wiring room

	I1	W1	W3	W4	W7	W8	W9	S1	S3
I1	0	2	1	2	3	4	4	2	5
W1	2	0	1	1	2	3	3	1	4
W3	1	1	0	1	2	3	3	1	4
W4	2	1	1	0	2	3	3	1	4
W7	3	2	2	2	0	1	1	1	2
W8	4	3	3	3	1	0	1	2	1
W9	4	3	3	3	1	1	0	2	1
S1	2	1	1	1	1	2	2	0	3
S3	5	4	4	4	2	1	1	3	0

1/14, 1/18, 1/18, 1/13, 1/24) = (.04, .06, .06, .06, .07, .06, .06, .08, .04). Solderman 1 is the most central.

EIGENVECTOR CENTRALITY

Eigenvector centrality is the refinement of degree centrality. As you remember, degree centrality is "blind" in the respect that it doesn't take into account anything past immediate friends. In many networks the importance of a node is dependent on the importance of the nodes to which it is connected, not the sheer number. In a high school an individual who is the friend of one very high status person becomes popular because of that one connection. The current men's basketball head coach at Oregon State University, Craig Robinson, enjoys a cache because he is President Obama's brother-in-law. In a communications network a node is exposed to lots of information if her immediate neighbors possess lots of information because they themselves are well connected.

An eigenvector **e** is a mathematical solution to this problem, the details of which we need not go into here. It is enough to say:

$$e_i \propto \sum_k a_{ik} e_k \text{ where " } \propto \text{ " means "proportional to."} \qquad (10.4)$$

The actual computation of eigenvector centrality is beyond the scope of this book, but the results are easy to illustrate. Consider, for example, the nine workers in the largest component of the bank wiring room, Figure 10.2. The centrality of each worker is proportional to the sum of the centralities of the actors to which he is connected. The measure is basically degree centrality, but it weighs each person's friends by the amount of friends they have, and then the amount of friends the friends of their friends have, and so on.

$$
3.22
\begin{pmatrix}
1.49 \\
4.43 \\
4.78 \\
4.43 \\
2.57 \\
1.61 \\
1.61 \\
5.04 \\
1
\end{pmatrix}
=
\begin{pmatrix}
4.78 \\
4.78 + 4.43 + 5.04 \\
1.78 + 4.43 + 4.43 + 5.04 \\
4.78 + 4.53 + 5.04 \\
5.04 + 1.61 + 1.61 \\
2.57 + 1 + 1.61 \\
2.57 + 1 + 1.61 \\
4.43 + 4.78 + 4.43 + 2.57 \\
1.61 + 1.61
\end{pmatrix}
=
\begin{pmatrix}
4.78 \\
14.25 \\
15.68 \\
14.25 \\
8.26 \\
5.18 \\
5.18 \\
16.21 \\
3.22
\end{pmatrix}
\qquad (10.5)
$$

In Figure 10.1, Solderman 1, who is connected only to other nodes of relatively high degree, is the most central by this measure.

BETWEENNESS CENTRALITY

Remember that centrality refers to two different phenomena: the degree to which a node in a network is advantageously placed and the degree to which a node is important for the functioning of the network as a means for distributing resources, where edges are transmission routes. Betweenness centrality is the latter type of centrality score.

Let's go back to your large retail store. Instead of placing your price scanner in a nice spot, suppose you have a new product that you want to promote. If you used the criterion represented through closeness centrality, you would place this product in the center of the store. However, the center of your store is not necessarily the place with the highest traffic, it's only the place that's the closest if someone from any random location in your store suddenly decides to confirm that you are really trying to hawk them an old Rick Astley CD for $15. What a rip-off. If you wanted to promote a product, you would probably want to place it in a location where most people would eventually pass by it. Any location by the checkout scanners or the entrance of your store would fit this new criterion.

In networks, betweenness centrality tries to find vertices that are in locations with high traffic. It represents a type of power that is distinct from the popularity of degree centrality or the influence of closeness centrality. For example, consider a network with three vertices: a factory, a wholesaler, and a retailer. The wholesaler's entire profitability is based on his betweenness centrality. If there were a direct connection from the factory or retailer, and no need to go through the wholesaler, the wholesaler would go out of business. That's the premise behind the

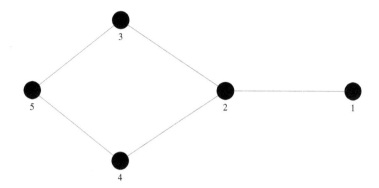

Figure 10.3. Example network for betweenness centrality.

marketing strategy of Costco, a wholesale retail store. Consumers believe they're getting better prices because they are bypassing the retail end of the chain, and buying wholesale. A newspaper editor has a lot of power because she is the gatekeeper between reporters and readers. She can selectively choose which stories to run and which stories to leave out. In social networks, betweenness centrality represents those who are in positions that offer unique advantages in the transmission or flow of goods or ideas.

Like closeness centrality, betweenness centrality is based on geodesics, or shortest paths. It measures the degree to which communication (or other quantities) flows through different vertices when members use the network to communicate with one another indirectly. To calculate betweenness centrality for a node i in a component of size n go through the following steps.

1. For each of the $(n-1)(n-2)/2$ pairs $\{j, k\}$ of vertices that do not include i, calculate all the geodesics connecting j and k that include and exclude vertex i. Take the proportion of geodesics connecting j and k that flow through node i.
2. Add up all these $(n-1)(n-2)/2$ proportions. This is the closeness centrality of vertex i. It's range is from 0 to $(n-1)(n-2)/2$.

To take a very simple example, look at the network in Figure 10.3.

Let us first consider Vertex 4. The following labeled matrix shows the proportion of the geodesics between every pair of vertices that pass through Vertex 4. For example, there is one geodesic between Vertices 3 and 4, and it does not pass through Vertex 4. Therefore the element in the row labeled 1 and the column labeled 3 is $0/1 = 0$. There are two geodesics between Vertices 1 and 5: 1-2-3-5 and 1-2-4-5. One of the two includes 4, so the element in the row labeled 1 and the column labeled 5 is 1/2.

$$
\begin{array}{c}
\quad\quad 1 \quad\ 2 \quad\ 3 \quad\ 5 \\
\begin{array}{c} 1 \\ 2 \\ 3 \\ 5 \end{array}
\begin{pmatrix}
X & \frac{0}{1} & \frac{0}{1} & \frac{1}{2} \\
X & X & \frac{0}{1} & \frac{1}{2} \\
X & X & X & \frac{0}{1} \\
X & X & X & X
\end{pmatrix}
\end{array}
\quad \text{for Vertex 4} \quad\quad (10.6)
$$

The sum of all the numbers in Equation 10.6 is $0+0+1/2+0+1/2+0 = 1$, the betweenness centrality of Vertex 4. The entries in the matrix follow the ordering of vertex numbers, omitting the vertex for which the score is given. Equations 10.7 and 10.8 show the calculations for the betweenness centralities of Vertices 2 and 5.

$$
\begin{array}{c}
\quad\quad 1 \quad\ 3 \quad\ 4 \quad\ 5 \\
\begin{array}{c} 1 \\ 3 \\ 4 \\ 5 \end{array}
\begin{pmatrix}
X & \frac{1}{1} & \frac{1}{1} & \frac{2}{2} \\
X & X & \frac{1}{2} & \frac{0}{1} \\
X & X & X & \frac{0}{1} \\
X & X & X & X
\end{pmatrix}
\end{array}
\quad \text{for Vertex 2} \quad\quad (10.7)
$$

$$
\begin{array}{c}
\quad\quad 1 \quad\ 2 \quad\ 3 \quad\ 4 \\
\begin{array}{c} 1 \\ 2 \\ 3 \\ 4 \end{array}
\begin{pmatrix}
X & \frac{0}{1} & \frac{0}{1} & \frac{0}{1} \\
X & X & \frac{0}{1} & \frac{0}{1} \\
X & X & X & \frac{1}{2} \\
X & X & X & X
\end{pmatrix}
\end{array}
\quad \text{for Vertex 5} \quad\quad (10.8)
$$

Adding up the values in the matrices for the different vertices, we find that the centrality scores for the five vertices is $(0, 3\frac{1}{2}, 1, 1, \frac{1}{2})$. Vertex 2 is by far the most central.

Looking at the bank wiring room, Solderman 1 in Figure 10.1 is "between" all the four nodes to his right and all the four vertices to his left. His betweenness centrality score is therefore 16. At the other extreme, Inspector 1 and Solderman 3 are not on any geodesics. The removal of a node that is high on betweenness centrality would eliminate many geodesics and therefore would tend to increase the distances between other pairs of nodes in the network. The removal of a node with a score of zero would have no effect on distances between other pairs.

Cut Points

Cut points are easy to define: the removal of a cut point increases the number of unconnected components in a graph, just like the elimination of an edge that is a bridge. In a communication network vertices that could communicate with each other indirectly cannot if the node is eliminated. Thus, cut points are a crude index of the importance of a node for the functioning of the network. In Figure 10.1 the removal Solderman 1 or Wireman 7 would divide the largest component in the bank wiring room into two almost equal parts. Wireman 3 is also a cut point because his removal would isolate on Inspector 1.

The following tables show all the measures of centrality for all the nodes in Figures 10.1 and 10.2.

At this point, please play around with the demonstration *Centrality* to become familiar with the properties of the measures. Pay particular attention to networks for which the different measures do not agree. The sizes of the vertices are proportional to their centralities using six centrality measures: degree, closeness, betweenness, cut points, eigenvector, and centers.

CENTRALIZATION

Centralization refers to the degree of inequality of centrality in a network. Centrality is a property of nodes, but centralization is a property of networks. Centralization is the degree to which centrality is monopolized by a small set of nodes in the network. If a network is centralized, there is a small core of highly central nodes and a large periphery of low centrality nodes. As we shall see later in the chapter on scale-free networks, the existence of a core within a network can have profound consequences for the operation of the network. Highly centralized networks can be more efficient in the distribution of information (once information reaches a core member it can easily and quickly be made available to all), but highly centralized networks can be very vulnerable if a core member is disabled. Inequality can also diminish the effectiveness of a network if participation by all members of a group is desirable; if all decisions are made by a small set of members, a group may not be able to take advantage of the knowledge and skills possessed by all its members. Highly centralized networks can also be very unequal in status and power, which may be considered undesirable on its own; a community in which only a small subset of interconnected individuals participate in all the governmental and nongovernmental organizations may be unhealthily and undemocratically centralized.

Once one has chosen an appropriate measure of centrality for the nodes, there is a standard way of computing the centralization of a network for that measure of centrality. The measure takes the star network (one central vertex connected to all others, none of which are connected to each other) as most centralized and measures the degree of deviation from that criterion. In the numerator of the measure of centralization is the difference between the centrality of the most central vertex and all other vertices. In

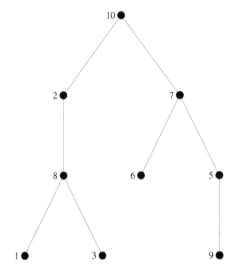

Figure 10.4. Network for Exercise 2.

the denominator is the sum of these differences in the star network with the same number of vertices, the most centralized structure. Suppose there are g actors in network G and actor $i*$ is the most central according to centrality measure A. Let $C_A(G)$ be the centralization of network G using measure A, and let $C_A(i)$ be the centrality of actor i using measure A. Then,

$$C_A(G) = \frac{\sum_j (C_A(i*) - C_A(j))}{\max \sum_j (C_A(i*) - C_A(j))} \qquad (10.9)$$

The "max" refers to the distribution of centrality scores in a star network of the same size. For example, consider two networks: G_1 is a line network A-B-C-D-E; G_2 is the complete network with five vertices. Let's calculate degree centralization C_D for these two networks. In the star network with five vertices the degrees are 4, 1, 1, 1, and 1. The value of the denominator of Equation 10.9 is $(4 - 4) + (4 - 1) + (4 - 1) + (4 - 1) + (4 - 1) = 12$.

Chapter Demonstrations

- *Centrality* illustrates various measures of network centrality applied to random networks
- *Centrality Game* tests your understanding of centrality by asking you to create networks with desired properties

Table 10.2.

Centrality scores for the bank wiring room network in Figure 10.1

	I1	W1	W3	W4	W7	W8	W9	S1	S3
Degree	1	3	4	3	3	3	3	4	2
Betweenness	0	0	7	0	15	3	3	16	0
Closeness	.043	.059	.062	.0591	.071	.056	.056	.077	.042
Graph center	0	0	0	0	1	0	0	1	0
Cut point	0	0	1	0	1	0	0	1	0
Eigenvector	1.49	4.43	4.78	4.43	2.57	1.61	1.61	5.04	1

Table 10.3.

Centrality scores for the marriage network in Figure 10.2

	Degree	Betweenness	Closeness	Center	Cut point	Eigenvector
Dall'Antella	1	0	.014	0	0	0.187
Ginori	3	35	.018	0	1	0.711
Dietisalvi	1	0	.014	0	0	0.187
Medici	5	44	.024	0	1	2.326
Guiciardini	1	0	.017	0	0	0.612
Tornabuoni	1	0	.017	0	0	0.612
Guasconi	6	73.5	.028	1	1	3.806
Albizzi	4	19.3	.024	1	0	3.092
Da Uzzano	1	0	.019	0	0	1.002
Rondinelli	2	9	.021	0	0	1.531
Della Casa	2	0	.020	0	0	1.531
Castellani	5	39.2	.020	0	1	2.008
Pepi	1	0	.015	0	0	0.529
Scambrilla	1	0	.015	0	0	0.529
Benizzi	0	4.5	.019	0	0	1.241
Peruzzi	6	51.3	.026	0	1	3.798
Strozzi	5	45.2	.026	0	1	2.703
Panciatichi	1	0	.017	0	0	0.712
Rucellai	1	0	.017	0	0	0.712
Aldobrandini	1	0	.018	0	0	1

EXERCISES

1. Using the demonstration *Centrality Game*, answer the question posed on the demonstration.
2. Using Figure 10.4, compute betweenness, closeness, and degree centralities for the following network. Hint about betweenness: In a tree there is only one geodesic between any pair of vertices.
3. Calculate closeness centralization for the centrality scores in Table 10.2.
4. Calculate betweenness centralization for the centrality scores in Table 10.3.

CHAPTER 11

Small-World Networks

Imagine that you are the supreme ruler of a number of islands within an archipelago. Your daily routine as dictator includes sipping on mojitos, getting a tan, jogging on the beach, and receiving wonderful massages. Occasionally, you do a few things that involve country management. Sometimes these things involve making hard choices. One recent choice you had to make concerns the allocation of boats on shipping routes.

Your country has only a limited number of boats. Either these boats can sail in and out of your country, docking at trading ports with the rest of the world, or the boats can be used to improve travel among the many islands that make up your country. Trading with foreign countries allows much needed resources to come in, such as food or other types of raw goods. In return, you export some things unique to your island, such as your very special coconuts. At the same time, if travel is made easier among your islands, it helps foster a sense of national identity and preserve your unique customs. Citizens within your country will now be able to travel to the island across the bay to visit their relatives, have a barbeque, and sing songs to your greatness with their extended families. This type of cohesion within your country is important—just in case a foreign country decides to invade you for your coconuts.

In the scenario presented above, the limited number of boats determines how you control physical space. When it comes to friendships and time, there are some parallels. Maintaining a good friendship may cost a considerable amount of investment and dedication on your part. It may be impossible to be good friends with everyone. Sometimes you have to spend quality one-on-one time with them to make them feel special, or invest money—another limited resource—on gifts or the cost of just hanging out.

In most countries, national borders are the result of politically drawn lines. Sometimes these lines are influenced by natural geographical features such as a river, a mountain range, or, in the case of your country, a sea. If we are to extend the analogy to friendships, the borders of your social circles are not as clear as a country's border. This is because the borders of your social circles are determined by whether or not your friends know each other. Suppose you met your main core group of friends from living on the same dorm floor as them. Then it is likely that some of your friends are friends with each other as well. Maintaining your social relations with this group of friends may be like investing in roads within your country. Investing in a road that leads outside your country may be like maintaining a relationship with a unique friend that is radically different from your core group of friends. This person also should not know any one from your core group of friends. So, in essence, your relationship with this person serves as a sort of bridge to a distant part of a social network.

Even this extended analogy is still not a clear-cut way to define the borders of your social circle. Sometimes you may have multiple groups or cliques of friends. Some friends may belong to more than one group. Even with these issues, the function of investing in relations within your group of friends versus relations with acquaintances on the fringe remains similar to building roads within and out of your country. You may receive unique social resources as a result of those fringe friendships. For example, your fringe friend is more likely to introduce you to something fresh and new than someone within your core group of friends, with whom you already share many of the same activities. This could be anything from a new restaurant recommendation to a new job offer. At the same time, you may enjoy spending time with your core group of friends, or at least find it easier, because of the redundant nature of within-group ties. Your core group of friends may all share a common reference point for gossip or have a common preference for music that prevents conflicts over what track is playing.

The bottom line is that our social connections function much like the roads in your imaginary country—they can be far-reaching or local. However, because of the interesting behavior of social network structure, our social connections can sometimes be both far-reaching and local. This is known as the small-world phenomenon.

SHORT NETWORK DISTANCES

In the previous chapter, we saw that random networks have very low distances between any two given vertices. The other network structures that were explored, such as lattices, did not have this basic property. As we will see, this property is important for social network models because evidence suggests that real-world networks tend to have very low distances as well. What this means is that our social connections function much like far-reaching roads—you don't have to keep changing highway routes to get where you want to go.

The Milgram Experiment

Social Psychologist Stanley Milgram was a classic social scientist who conducted a number of well-known experiments. You may have heard about his obedience experiments, where subjects were goaded into applying seemingly painful electric shocks to victims that were really actors. The grim black and white photos from the experiment would work as well in a horror documentary film as they would in a social psychology textbook.

In 1967, Milgram (Travers and Milgram, 1969) conducted another important experiment that was perhaps less well-known because it wasn't as shocking—excuse the pun. To put it in network terms, Milgram was interested in exactly how short distances were among individuals. Since he wanted results to be as surprising as possible, he chose subjects who were considered socially distant from each other. Thus instead of picking two random individuals in the same city, he choose a target individual in Boston, Massachusetts, a fairly urbanized East Coast city, and then selected subjects in Omaha, Nebraska, a more remote town in the Midwest. A subject chosen in this way would probably have different musical tastes than the target individual living in the city, or even a different accent. Keep in mind that that the Internet and cell phones didn't exist back then, so the social space between those two cities was a bit wider.

The goal of the study was for the participant in the Midwest to try to reach the target person through a chain of letters sent by mail. Each person in the chain could forward the letter only to a person whom they knew on a first-name basis. Of the 296 letter chains initiated by the Midwest residents, only 64 eventually reached their target resident in Boston. If you think about it in today's terms, this was an amazing result. Some of us get fairly annoyed at even moving our wrists and clicking our fingers to delete the spam in our email inbox. Imagine receiving one of these chain letters in 1964, and then having to write on paper, the name and address of someone you might not even know that well, and then having to get up to put this annoying packet in a mailbox. What a chore!

However, among the 64 letters that did reach their destination, it took an average of 5.5 steps. This result eventually was developed into the popular phrase "six degrees of separation", or the idea that we are all somehow connected by six steps on average. Milgram's experiment not only suggested that real-world social networks had a low average distance but also demonstrated various methods that individuals used to search their networks. The subjects sent letters to contacts whom they knew who lived closer geographically to the target person. Of course, the path that the chain letter took to get to the target might not necessarily have been the shortest path. There might have been contacts whom the participants knew who might have been much closer to the target. It is also entirely possible that participants ended up forwarding the packet to an acquaintance who was even further from the target person than they were. Nonetheless, even without a map of the web of social relations or tempting incentives, Milgram demonstrated that people could reasonably navigate a packet through a large complex network.

Milgram's classic experiment was recently replicated on a worldwide scale by a research team led by Duncan J. Watts (Dodds et al., 2003) at Columbia University. Instead of document packets, participants used email. The results indicate that Milgram's initial ballpark of 5.5 was pretty close, as the average length of the completed email chains was about 5.

Kevin Bacon: Center of the Universe?

In April 1994, someone made a claim on an Internet newsgroup, which was the equivalent of the modern-day Internet forum back then, that a movie actor named Kevin Bacon was the center of the universe. The poster wasn't trolling. As proof, he proposed that any movie actor, no matter how obscure, could be linked back to Kevin Bacon in a very few number of steps. Movie actors were considered connected if they starred in the same movie together. The post quickly grew in length with other users putting the claim to the test. Eventually this turned into a popular game for those with an unhealthy knowledge of film actors. Someone nominates an actor, and another person tries to find the shortest distance between that actor and Kevin Bacon. This distance became known as the Bacon Number.

Today, there's a website called the Oracle of Kevin Bacon, where users can input any actor, and using the Internet Movie Database (IMDb), the website will find the shortest path between the given actor and Kevin Bacon. A bookmark to this website on your Internet phone is quite useful if you ever need to impress your friends at a party who happen to be playing the Kevin Bacon game—providing that you access the website discreetly.

Kevin Bacon isn't necessarily the center of the universe. He is, however, very well connected from starring in a wide variety of movies in different genres over a long period of time. And even though there are other actors who have an even shorter average distance to every other actor, the difference in distance between them and everyone else and some obscure actor and everyone else isn't that great either. The moral of the story is, however, that distances tend to be short in general, which is consistent with the results of Milgram's experiment.

SOCIAL CLUSTERING

At the beginning of this chapter, connections within social groups were used as an analogy for local roads. We now introduce a measure for quantifying how local or clustered a particular individual is in a network.

Density

As an abstract concept, density is simply how full or packed an object is. Suppose we convert this into a measure that is bounded by 0 and 1. You

Figure 11.1. Network with two isolates.

could think of an empty glass of water being measured as 0, a glass half full (or half empty, for all you pessimists out there!) of water being measured as 0.5, and a glass filled to the brim being measured as 1. For networks, density describes how many edges exist over the total number of possible edges. If a graph is completely empty, it will have a density of 0, and if the graph is a complete graph, it will have a density of 1.

Suppose we have a network with only two individuals, and the relationship represented by the network is bidirectional friendship. Then these two individuals can be either connected or not, and therefore there are only two possible density values—0 and 1. If the relationship represented in this network were a directional relation such as respect, then there would be three possible density values: $\{0, 1/2, 1\}$. These individuals can both not respect each other, or one can respect the other without reciprocation, or they can both respect each other. Therefore, in directed graphs, the arcs are counted as the relationship in the network, and represented by m, whereas in undirected graphs, edges are represented by m. Because there are twice the total number of possible relationships in directed graphs, it is important to remember whether you are dealing with undirected or directed graphs when calculating density. We now generalize the formula for density in networks.

For undirected graphs, the density is given by,

$$d_{undirected}(G) = \frac{m}{\frac{n(n-1)}{2}} = \frac{2m}{n(n-1)} \qquad (11.1)$$

where m is the number of edges and n is the number of vertices in the graph. The numerator is simply the number of relationships in the network, and the denominator represents the number of ways to select unordered subsets of 2 from a set of size n. In other words, the denominator represents all the ways to pick pairs of two people from the list of people in your network. For directed graphs, the formula simply becomes,

$$d_{directed}(G) = \frac{m}{n(n-1)} \qquad (11.2)$$

Example 1. What is the density of the network given in Figure 11.1?

The network in Figure 11.1 has 12 vertices (including the 2 isolates) and 9 edges. Using the formula for the density of an undirected network, we get,

$$d_{undirected}(G) = \frac{2m}{n(n-1)} = \frac{2(9)}{12(11)} = 0.136 \tag{11.3}$$

Alternate Conceptualization

Another method of interpreting both the concept and the formula of density is to envision the adjacency matrix of a network. Recall that the adjacency matrix of a network is a $n \times n$ matrix where n is the number of people in the network. The entry in row i, column j in the matrix represents the value of the relation between person i and person j. In most types of networks, this is merely a binary value that reflects whether the tie is present (1) or not (0). Recall that the diagonal of an adjacency matrix is meaningless, as the people in the matrix cannot be connected with themselves. Therefore the number of possible valid cells in directed networks is $n^2 - n$, where n^2 represents the total number of cells in the matrix and $-n$ excludes the diagonal. Factoring out n, we obtain $n(n-1)$.

Example 2. Given the matrix $\begin{pmatrix} 0 & 1 & 1 \\ 1 & 0 & 1 \\ 1 & 1 & 0 \end{pmatrix}$, the density is 1.

In Example 2, all of the nondiagonal cells in the matrix are filled, so the network represented by the matrix has a density of 1.

Example 3. Given the matrix $\begin{pmatrix} 0 & 1 & 0 \\ 1 & 0 & 1 \\ 0 & 1 & 0 \end{pmatrix}$, the density is 4/6 or 2/3.

If the matrix given in Example 3 represented a digraph, then four out of the six possible nondiagonal cells are filled, so the network has a density of 4/6. Since the matrix is symmetric, with cell m_{ij} being equivalent to its reflection across the diagonal, cell m_{ji}, the matrix can also be seen as representing an undirected graph. In this case, 2 out of the 3 possible nondiagonal cells are filled.

Ego Networks

An ego network is sort of a mini network that consists of just an individual and the other individuals whom she is connected to. For example, your ego network would contain yourself and your immediate friends, and the relationships among you and your friends. However, it would not contain a friend of your friend whom you were not friends with. The diameter of an ego network can be at most two, for example, when you have two friends who do not know each other. A normal network of n individuals has n distinct ego networks.

Although an ego network is lacking in the rich structural features that normal, full-sized networks have, it is still popular and widely used in research because of its simplicity. One problem with collecting data on networks is defining the boundaries of the network. Suppose that you were not collecting ego network information, but instead you were mapping out the entire friendship network in a population. You ask one respondent who his friends are, and then you ask those friends who their friends are. Each new respondent refers you to her friends, and pretty soon the number of respondents in your study snowballs out of control. It is difficult to come up with a nonarbitrary criterion for when to stop collecting data. With an ego network, the boundary is determined by definition. You simply interview one respondent, then interview their friends, and stop.

Additionally, social networks that represent a subset of a larger population are likely to be unrepresentative of the population as a whole. Because of the principle of homophily, those in the same network are likely to share the same traits or attributes. Because ego networks are centered around a specific individual, as long as the person is representative, the set of ego networks may be somewhat representative. Some large-scale random surveys, such as the General Social Survey (GSS), have specific years when information on ego networks is collected.

Clustering Coefficient: Local Density

Now we combine the concepts of both network density and ego networks and arrive at a clustering coefficient. The clustering coefficient of an individual is just the density of that individual's ego network *with the individual removed*. In other words, it's the density of the network formed by the individual's friends and the relations among them. The last part about excluding the original individual is important because clustering coefficients are not simply the density of the ego network. The individual and ties from the individual to her friends are excluded because that information is trivial and redundant by definition. Everyone has a friendship relation with his friend. If he does not have a friendship relation with the friend, then that person is not a friend and is excluded from the ego network.

Example 4. Suppose Adam has three friends: Bob, Conrad, and Dan. Of Adam's three friends, only Conrad and Dan are friends with each other. What is Adam's clustering coefficient?

The correct way to calculate the clustering coefficient is to consider the network of only Adam's friends, which consists of Bob, Conrad, and Dan ($n = 3$). Since only one friendship exists in this network, the tie between

Conrad and Dan, $m = 1$. Using the formula for the density of an undirected network, we get,

$$d_{undirected}(G) = \frac{2m}{n(n-1)} = \frac{2(1)}{3(2)} = 1/3 \qquad (11.4)$$

Thus, the clustering coefficient for Adam is 1/3, which is correct.

Now, suppose we forgot to exclude Adam and the ties to his friends in the calculation and computed the density of Adam's ego network. This network would have four people ($n = 4$), and it would have four edges: Adam's three ties to his three friends and the extra tie between Conrad and Dan ($m = 4$). If we used the formula for the density of an undirected network in this situation, we would get,

$$d_{undirected}(G) = \frac{2m}{n(n-1)} = \frac{2(4)}{4(3)} = 2/3 \qquad (11.5)$$

The above answer of 2/3 is not the correct clustering coefficient for Adam, but it represents the raw density of Adam's ego network. Thus, we see that although the clustering coefficient can conceptually be considered the local density, or the density of an actor's ego network, it is not exactly the same thing in practice.

It is important to note that the clustering coefficient maps an individual in the network to a number between 0 and 1. It is not a measure associated with the entire network. However, when we talk about the clustering of a network, we are talking about the average of the valid clustering coefficients for everyone in the network. Individuals who are disconnected and isolated are not included in the calculations because their friendship network does not exist. Individuals with only one connection are also excluded because their clustering coefficient is undefined. If they have only one connection, then the denominator of their clustering coefficient is $n(n-1) = 1(0) = 0$, giving you a divide by zero scenario. Therefore, the clustering of a network measure is not a real average of clustering coefficients for all individuals in the network, but only for those with two or more connections.

Sometimes the terms "clustering coefficient" and "clustering" are used interchangeably for individuals and networks. For the purposes of this text, we will refer to clustering coefficients as a measure of the local density of a specific individual, while clustering will refer to the average of all valid clustering coefficient values in a network.

Another way to interpret the clustering coefficient of an individual in a friendship network is to think of it as the probability that the individual's friends are friends with each other. Someone with a high clustering coefficient probably spends a lot of time in group activities with her friends. For example, they could all be students who live in the same dorm or frat house or play on the same sports team. The group activity

Table 11.1.

How density decreases in large networks

n	$m = 15n$	$d_{example}(G) = \frac{2m}{n(n-1)}$
31	465	1
100	1,500	0.303030
500	7,500	0.060120
1,000	15,000	0.030030
10,000	150,000	0.003000
1,000,000	15,000,000	0.000030
6,700,000,000	100,500,000,000	0.000000

increases the likelihood that everyone knows each other. Someone with a low clustering coefficient might be someone who has a lot of Internet friends from different places who may not necessarily know each other.

Clustering in Random Networks

In random networks, the clustering of the network is approximately equal to the density of the network. If on a macro level, the entire random network is dense, then on a micro level, when you examine the local network around a single individual, that network is likely to be dense. Conversely, if the random network is sparse overall, then the local networks around individuals will be sparse as well.

If we use random networks as our model for real-world networks and assume that people could maintain only a finite number of relationships, then in very large populations, the density of such a random network approaches zero. For example, let's make the reasonable assumption that, on average, someone maintains 30 friendships. Since each friendship tie consists of two people, then in this network, each person will contribute 15 to the number of edges ($m = 15n$). Thus the formula for the density of this network is,

$$d_{example}(G) = \frac{2m}{n(n-1)} = \frac{(2)(15n)}{n(n-1)} \tag{11.6}$$

Table 11.1 gives the density of the network with different numbers of people. When this network has 31 people, the density of the network is 1 because each person is connected to all the remaining 30 people in the network, and thus all the possible edges exist. Notice that as the number of people in the network increases, the density of the network decreases. If we used this random network to model the approximately 6.7 billion people on earth, the density would be practically zero.

Since the network is random, this means the clustering is practically zero as well. This is because we know in random networks, the density of local networks on the individual resembles the entire network. This means that if friendship networks were random, it would be almost statistically impossible for your friends to be friends with each other. However, we know this is not the case. Although random networks exhibit the low path lengths that we observe in real-world networks, they do not exhibit the clustering property. We now introduce the small-world network model, which combines both low path lengths and relatively high clustering.

THE SMALL-WORLD NETWORK MODEL

Watts and Strogatz (1998) presented a network model that was both far-reaching and local. The idea is elegant in its simplicity: start with a localized graph with high clustering and high average distances between vertices, and then randomly rewire edges. Rewiring is defined as deleting an existing edge and creating a new edge between two previously unconnected vertices. With each rewired edge, the expected distance between any two vertices decreases. The clustering coefficients, which were initially high, also decrease. However, an interesting side effect of the rewiring is that the average distance between vertices drops much faster relative to clustering. The average distance between vertices in the rewired networks approaches that of a random graph with only a few rewires, while clustering can still remain fairly high. When a network is rewired to a point where it exhibits these properties, it is considered a small-world network. However, with additional rewiring, the clustering will continue to fall until both distances and clustering are low. The resulting network at this point is no longer a small-world network but a random one.

In addition to introducing the small-world network model, Watts and Strogatz suggested that this small-world network structure was natural. While it may have been created due to random chance, there was something unique and special about it. In addition to the social example of the movie actor network, Watts and Strogatz showed that networks with high clustering and low average path length also occurred in the neural network of an earthworm and also in power grid networks. Later on, this structure was discovered in many more natural phenomena by other researchers. The small-world network therefore has broad interdisciplinary appeal.

We now explore a few examples of the generation of a small-world network.

Clustered Circle Graph

The clustered circle graph was one of the first models that Watts and Strogatz used. They called it the α model. Figure 11.2 shows a clustered circle graph with 10 vertices. With this small graph, you can notice that

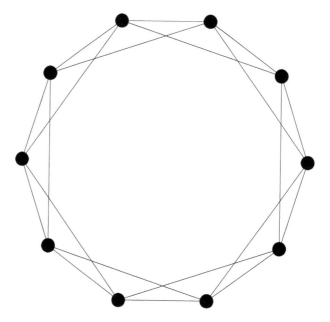

Figure 11.2. Clustered circle graph, $n = 10$.

there is a pattern to how each vertex is connected: each vertex in this graph is connected to four other vertices, the two closest neighbors in both the clockwise and counterclockwise direction. Suppose we had a clustered circle graph with 50 vertices. First, let's find the density of this graph. We know that $n = 50$, and we want to obtain the number of edges in this graph, m. We know that each vertex has 4 edges, so the average degree is 4. Using the formula for average degree, we deduce that the network has 100 edges.

$$avgdeg(v) = \frac{\sum deg(v)}{n} = \frac{2m}{n}$$

$$4 = \frac{2m}{50}$$

$$200 = 2m \tag{11.7}$$

$$100 = m$$

Recall that the local density or clustering for a random network is approximately equivalent to its overall density. So, if this were a random graph, we would expect a clustering coefficient of 0.166667.

$$\frac{2m}{n(n-1)} = \frac{2(100)}{50(49)} = 0.082 \tag{11.8}$$

However, because the graph isn't random, let's calculate the initial clustering coefficient for our clustered circle graph. Since all vertices are in an identical position, we need to calculate the clustering coefficient for only one vertex, and assume that value to be the average of all positions. Let's pick a random vertex, and call its neighbors A, B, C, and D. Among these four vertices there are three connections: {A–B, B–C, C–D}. Thus $n = 4$ and $m = 3$. Using the density formula, we get,

$$\frac{2m}{n(n-1)} = \frac{2(3)}{4(3)} = 0.5 \qquad (11.9)$$

As 0.5 is much greater than 0.082, we see that this network has a much higher than expected value for clustering in a random graph.

Another thing to note is that distances in this clustered circle graph are relatively high. The diameter of this circle graph is 13. If you picked a pair of vertices that were spatially the furthest from each other, it would take 13 hops to reach the other. You'll have to trust me on the next calculation, but the average distance between all 1,225 pairs of vertices is 6.63. This is a huge distance considering there are only 50 vertices in the network. Remember that according to Stanley Milgram, and later on Duncan Watts, in a world with more than 6 billion people, the proposed average distance between any two random individuals is only about 5 or 6!

Figure 11.3 shows the network after one rewiring and then five rewirings. Notice that dramatic changes can already be seen in the network after just one edge is randomly deleted, then readded. The new edge forms a sort of bridge in an area that is fairly close to the center of the network. If you had to bet some cash on it, you would probably bet that the average path length in the network has decreased dramatically from this new connection. To be exact, the average distance has decreased from 6.63 to 5.87. Because the rewiring is a random procedure, if we reset the network and randomly rewired another edge, the change could be larger or smaller depending where the new bridge forms. The best decrease in distance would be if the new edge appeared exactly in the center of the circle graph, dividing it evenly into two smaller ovals. After five rewirings, the average distance has dropped to 4.83. On the other hand, after five rewirings, the clustering of the network is still around 0.39.

While the small-world model may at first appear abstract and mechanical, without analogues in the real world, it does offer some sociological insights that may be worthy of discussion for those who like to sit at cafes with their goatees, shades, and Frappuccinos. Duncan Watts compared the early clustered start of a small-world network model to a primitive caveman society, where humans lived as isolated tribes. Watts also argued that the random network, which is the ultimate product of the prolonged rewiring process, could be representative of a futuristic society, where someone's set of friends may include random people spread across the entire planet because of communication and travel technologies. Our real world could be thought of as somewhere in the middle.

The idea is not entirely new. Social theorists have often pondered the effects of modernization and the role of increasing distant connections.

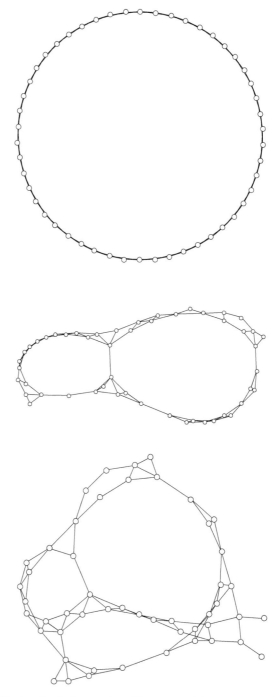

Figure 11.3. Clustered circle graph with no rewiring, one rewiring, and five rewirings, $n = 50$.

Anthony Giddens, one of the most influential sociologists of the 20th century, argued in his theory of structuration that as societies become modernized, the concept of time becomes increasingly independent from space, and subsequently the concept of space becomes independent from place. In a society without wheels, lightbulbs, or a sundial, an individual's daily routine and environment are very dependent on each other. Consequently, there will be a lot of overlap with other individuals in the same community, which could lead to what we call high clustering. Today, clustering still exists in our social networks but is just as often caused by similarity of interests or hobbies, as the space in which someone inhabits. Humans have increasingly become more mobile over the ages, through horses, the wheel, the boat, and the plane. The tools for communication have increasingly defied space through the telephone, the cell phone, the Internet, and now Internet cell phones. Communication tools have also become independent from time through writing, the phonograph, the cassette, the compact disc, and now Internet streaming videos. These have increasingly allowed the people greater opportunity to form random, far-reaching connections. The impact is that someone's local social space is increasingly "penetrated by and shaped in terms of social influences quite distance from them" (Giddens, 1990, 19). Giddens presents a number of consequences for such a phenomenon. For example, the property for social structures to be both local and global facilitates the growth of rationalized bureaucratic organizations. Thus, the small-world model could be considered an abstraction for the effects of modernity on social structure.

Even if you don't buy the story above, there is also the idea that the small-world concept through social network analysis has turned something that might have previously been thought of as abstract, qualitative, and unmeasurable into something concrete and quantitative. Network diagrams essentially show social structure, typically an abstract concept before social network analysis. We can implement the abstract terms coined by Giddens, such as "phantasmagoric" or "distanciation" as the average path length. Social cohesion could be operationalized as clustering coefficients, although there are many other ways to define and measure the concept of cohesion. We can collect network data on two separate societies, compare a network statistic such as clustering, and not only say which one has more clustering but also calculate how much more. Being able to say that the speed of light is 299,792,458 meters per second is much more precise than saying that the speed of light is very, very fast. Of course, many abstract sociological concepts may lose a lot of color when reduced to mere numbers, but measures can always be refined.

Finally, remember that even though models may not be empirical, they are at least theoretical. The random network model was a point of comparison, something to which we can compare our empirical networks and point out differences. At the very least, models may serve as our null hypothesis or controls. With the small-network model, we have noted the discrepancy in clustering between random and many real-world social networks, and have adjusted for it. In the next chapter, we explore another discrepancy.

EXERCISES

1. Given a clustered circle graph with 10 vertices and 4 connections per vertex, calculate the expected clustering given a random graph with the same number of vertices and edges, and the actual clustering of the circle graph. What is the diameter of this graph?
2. Given a 10×8 two dimensional lattice with 80 vertices, with each vertex being connected to its immediate orthogonal neighbors, calculate the expected clustering given a random graph with the same number of vertices. Next, calculate the actual clustering of the lattice. What is the diameter of this graph?

CHAPTER 12

Scale-Free Networks

Breaking news: In the fashion world of statistical distributions, the bell curve of the normal distribution is no longer hot nor fashionable. Indeed, reports are coming in from Milan, Paris, London, and other statistical fashion centers that pink is the new red, small is the new big, and the power-law distribution is the new normal distribution. Experts say that the image of the bell curve is forever tarnished in the minds of college students by negative associations with stressful college entrance examinations, harsh grading practices, and even the title of an unpopular book.

Okay, so maybe we exaggerate a bit, as the bell curve and its relatives, the binomial, Poisson, and t-distribution, are still very useful in keeping the engines of modern statistical analysis running smoothly. Also, there aren't really such things as statistical fashion centers, but you probably already had a hunch that was the case.

But the power-law distribution has recently attracted much attention, especially in emerging areas of study such as network and complexity science. One remarkable aspect of this distribution is that it tends to show up in many complex systems, even if these systems are very different from each other. For example, it describes the distribution of the population size of cities, the magnitude of earthquakes, price movement on stocks, and the occurrence of words in written human language. What's of even more interest to us is how the power-law shape describes the distribution of degree for many different types of networks, such as power grids, citations in scholarly journals, associations among movie stars, and the structure of Internet links. If it didn't cost anything to maintain actual human relationships, a lot of social networks would come pretty close to fitting a power-law distribution.

Albert-László Barabási and Reka Albert (Albert and Barabási, 2002) coined the term "scale-free network" to describe networks that exhibit power-law distributions. In this chapter, we will describe the power-law distribution, show how scale-free networks can be generated with such a distribution, and give some properties of scale-free networks.

POWER-LAW DISTRIBUTION

Let's suppose for a moment that the amount that individuals could earn each year averaged around $40,000. Let's also suppose that the distribution of income followed a bell-shaped or normal distribution. If we assume these two things, life would seem a lot more fair. While it is true that some people will inevitably make more than others, we would never see the levels of inequality we see today. With a normal distribution, it would be a statistical impossibility for anyone to earn an amount that is too far from the average income. There would be no such thing as billionaires or even millionaires. These high amounts of income would just be too many standard deviations away from the average. If income were really distributed this way, there would be a lot less discontent about inequality, and probably a great deal fewer scientists devoted to studying social stratification and inequality.

There is a particular way that income and wealth are distributed in our society that may seem unfair. There are some individuals who earn a hundred or even a thousand times more than what the mass majority of others earn. Their income allows some of them to indulge in lavish spending habits that have been visible for quite some time. Even in the 19th century, it drove economist and sociologist Thorstein Veblen (Veblen, 1965) to coin the term "conspicuous consumption" to describe the tendency for some of the super wealthy to flaunt their wealth. Therefore the distribution of income or wealth in society is nowhere near a bell-shaped or normal distribution. We need a different type of curve to show that while most of us drive around in a dented old Toyota, there are some who cruise around in a Ferrari.

Figure 12.1 shows the relationship between productivity and time to deadline for a typical college student. The x-axis represents the time left until the due date for a paper or project, and the y-axis measures how much work will be done at that time.

Just kidding. Actually, Figure 12.1 shows the general form of the Pareto distribution, coined after the Italian economist Vilfredo Pareto. The x-axis is the value of the variable, and the y-axis is the frequency in which that value occurs. Notice that this distribution is heavily right skewed. Unlike a normal distribution, there is an effective non zero probability for the existence of values much greater than the majority of lower values. In the top half of Figure 12.1, the entire range of values for the x-axis is shown. Notice that some values are so high that the majority of other values appear as a steep cliff on the left side of the histogram. The bottom half of the figure shows the histogram of the same distribution, but omitting all values above five. This close-up view let's you see the shape of the power-law curve up close. The power law has many forms and many names, and the Pareto distribution is but one type of power law. A discrete version of this distribution is called the Zipf distribution, which was originally used to describe the distribution of the word

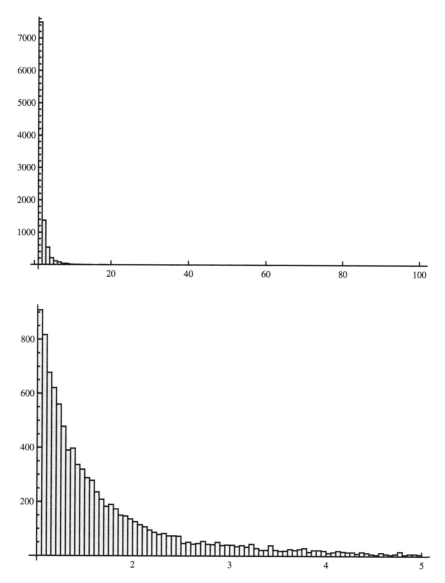

Figure 12.1. Top: A generic power-law (Pareto) distribution ($N = 10,000$). Bottom: The same distribution, showing only values between 1 and 5.

frequency in a language. We refer to the general class of all these distributions as power-law distributions.

One characteristic of power-law distributions is that the mean is much higher than the median value. What this means is that, in the case of describing income, most people earn an amount that is below average. This is because the income of the super rich artificially pulls the average

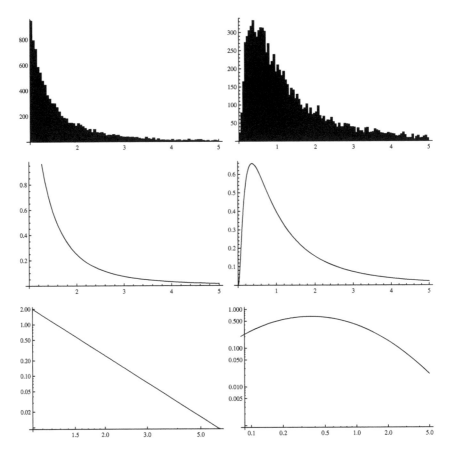

Figure 12.2. Randomly generated numbers from a power-law (left) and a log-normal (right) distribution.

income of everyone upward. When distributions are so skewed, a standard practice for statisticians is to prefer the median over the mean as a measure of central tendency.

Fake Power-Law Distributions

Because of the wild popularity of this trendy power-law form, many aspiring distributions often fool the researcher into thinking they are power-law distributions. But, never fear, we realize that you will accept no imitations, so we will demonstrate how to sniff out these counterfeit distributions masquerading as power laws.

Most of us are accustomed to distribution data being presented to us in some type of histogram form. To the naked eye, it's often fairly hard to discern a true power-law distribution from other distributions that have a similar form. Figure 12.2 has two columns; the column on the left shows a randomly generated power-law distribution, while the column on the right shows a randomly generated log-normal distribution. If you look only at the histograms in the top row, the only difference seems to be that the log-normal distribution starts off with low values, then climbs up steeply in the beginning. If you ignore the first few values along the x-axis, the two distributions would seem nearly identical.

The middle row gives a line plot of the probability density function of both distributions. If you sampled repeatedly from the distribution, up to a near infinite amount of times, the closer the histogram will come to resembling the shape of the probability density function. Again, if you ignore the initial values of the two distributions, the shape of their curves would look strikingly similar.

Finally, the third row shows the log-log plot of both distributions. The meanings of both the X- and the Y-axes remain the same as a histogram. The X-axis shows the value of the data point, while the Y-axis represents the number of data points that occur at that value. But the major difference is that the scale of the X- and Y-axes has changed. Instead of having distance being measured at even intervals, each axis is on a logarithmic scale, much like the well-known Richter scale that is used to measured earthquakes. The closer the distribution is to the power-law form, the more linear the trend line will be that would describe the data points. From Figure 12.2, you can see that something is beginning to look fishy about our power-law imposter on the right. It doesn't look as straight as the power-law distribution. Instead, the line is a bit curvilinear, with a slight arch shape. Thus, we can see that under the scrutiny of a log-log plot, a real scale-free distribution will appear as a nice straight line. Even though the log-normal distribution appeared very similar from the histogram, ultimately the log-log plot exposed it as a cheap rip off.

PREFERENTIAL ATTACHMENT

Recall that a basic small-world network can be created by randomly rewiring edges in a highly clustered network. The process of rewiring itself doesn't define a small-world network. In fact, remember that if too many edges are rewired, the network eventually becomes a random network. The interesting aspect of a scale-free network is that it comes closely packaged with a process that generates it. This process is called preferential attachment, or the tendency for nodes to prefer more highly connected nodes when forming new connections. This simple concept forms the basis of the Barabási-Albert (2002) model for generating scale-free networks, which we discuss here.

You first start with an initial network where each node has at least one connection. If a node initially has zero connections, then it has a zero chance of getting any new connections under this model, and thus it will live a life of solitude and die lonely and unknown. As time progresses, you simply add more nodes and connect them to existing nodes, with a preference for the more highly connected ones. Unlike random rewiring for small-world networks, the longer the preferential attachment mechanism is allowed to continue, the closer the network becomes to a pure scale-free network.

Preferential attachment is the tendency for new nodes in a network to prefer forming connections with popular existing nodes. Note that this does not mean that a new node will choose the most popular node every time, only that it is more likely to choose popular nodes. In the most common case, the probability that Node A chooses Node B is directly proportional to Node B's number of friends.

For example, consider that we have a friendship network with two nodes, Adam and Bob, each friends with each other. If a third person, Colleen, enters the network and chooses to form a friendship connection with Adam and Bob with equal probability, Colleen has a 50/50 chance of choosing Adam or Bob. If preferential attachment operates in this case, and Colleen befriends Adam and Bob with probability proportional to their friendship, there is still a 50/50 chance that Colleen chooses Adam over Bob or vice versa. Random choice is the same as preferential attachment in this case because both Adam and Bob have the same degree.

Suppose now that Colleen chooses to befriend Bob. We now have a network Adam-Bob-Colleen, with Bob being in the center with two connections. In this case, if a fourth person, Dan, enters the network and chooses to form a *random* connection with an existing person, there is a 1/3 chance that it will choose any existing person. However, Dan chooses to form a connection based on preferential attachment, then there is a higher likelihood that Dan will choose Bob, since Bob is the most popular person with two friends. Here, Dan has a 1/4 chance of connecting to Adam or Colleen, and 2/4 or 1/2 chance of connecting to Bob.

Let's make sure we've got this down. Suppose Dan ends up choosing Bob as a friend. Then we introduce another person, Esther, and she befriends Dan. Table 12.1 shows the network that describes our story so far. Now Freddy is entering the network, and he must decide whom to befriend. If Freddy were to pick a friendship randomly, then he would have a 1/5 chance of picking any of the existing people, since everyone would be picked equally under random chance, and there are five existing people. However, if Freddy were to choose his friendship based on preferential attachment, then he would have a 1/8 chance of picking Adam, Colleen, and Esther, 3/8 chance of picking Bob, and 2/8 chance of picking Dan.

Table 12.1.
Attachment probabilities under different systems for Freddy

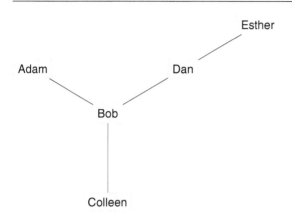

| | **Existing Network** | | | | | |
	Adam	**Bob**	**Colleen**	**Dan**	**Esther**	**Total**
Random attachment	1/5	1/5	1/5	1/5	1/5	1
Degree of node	1	3	1	2	1	8
Preferential attachment	1/8	3/8	1/8	2/8	1/8	1

Notice that the numerator in the probabilities is simply the degree of the node and the denominator is the total sum of degrees.

In general, if preferential attachment operates, the probability of a new node y entering into a network and connecting to node x is given by the following equation (where m is the number of edges):

$$\Pr(y \Rightarrow x) = \frac{Deg(x)}{\sum_{i \in V} Deg(i)} = \frac{Deg(x)}{2m} \qquad (12.1)$$

This is simply the degree of x, over the total sum of all degrees.

In general if we add almost a thousand more nodes to the network via preferential attachment, we end up with a network that looks something like Figure 12.1. Because of the sheer number of nodes in the network, only the connections are drawn for clarity. The highly connected nodes in the center are likely to be the original nodes, and nodes on the outer rims of the circle are likely to be fresh new nodes. New nodes have a higher chance of connecting to one of the original center nodes, and a lower chance of connecting to a specific node on the rim. However, because of the sheer number of nodes on the rims, sometimes new nodes will connect to nodes with few connections.

The network generated by preferential attachment in 12.3 looks very artificial. You probably won't find many social networks that have a similar

Figure 12.3. Scale-free network generated by preferential attachment ($N = 1,000$).

visual structure. However, the network represents only the most basic type of scale-free network. The important thing is that it generates a power-law distribution of degree, something that neither small-world nor random models will do. If the process of preferential attachment were to occur in social networks, it would represent a world where the only criterion that people used to select a friend was the friend's popularity. While that doesn't sound too far-fetched, keep in the mind that the model presented above allows each new person to choose only one friend, and that the

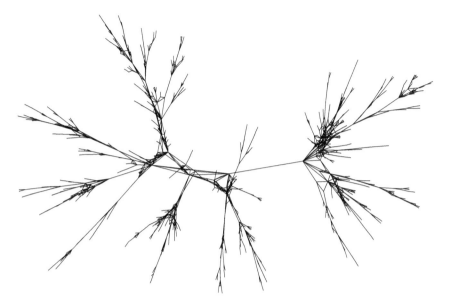

Figure 12.4. Scale-free network generated by the copy model ($N = 1,000$).

most popular people are popular because they're old, and not necessarily because of some other trait.

There are conditions in the real world in which preferential attachment may occur. Adolescents entering a high school may be most attracted to kids who are already popular. Websites may link to already popular and well-known websites. New immigrants to a country may be attracted to large cities with lots of jobs. However, we will now briefly explore other methods of generating scale-free networks. The thing that ties them all together is that a scale-free network must have a power-law degree distribution. Other than that, there is considerable freedom on how to create a scale-free network. While models don't necessarily have to be a perfect representation of what is happening in the real world, it would be nice if the process that generates the model has some sort of resemblance to a real-world process.

Alternative Scale-Free Network: The Copy Model

Let's examine the copy model (Kumar et al., 2000). Although this model was originally used to describe the structure of the Internet, it could just as easily be modified to describe social processes as well. The way it works is simple: You start with a small network, let's say with just two people connected to each other. At each new step, a new node enters the network and forms a random connection with an existing node. The new node also

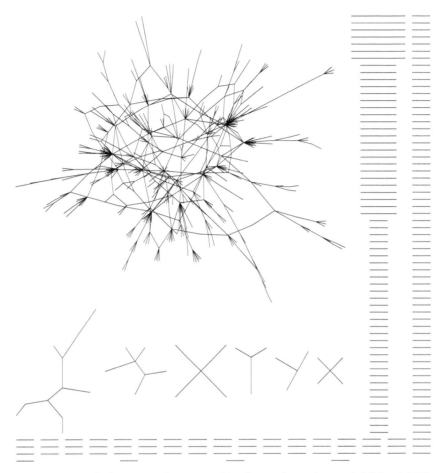

Figure 12.5. Scale-free network generated by the configuration model ($N = 1,000$).

forms connections with a random sample of the existing node's friends. If you repeatedly add new nodes and repeat this process, you will end up with a scale-free network. Like the preferential attachment model, the copy model is a generative model, which means it is formed by growing a network gradually by introducing new nodes, and connecting them to existing nodes. However, it is not like preferential attachment because new nodes that enter the network choose their connections randomly.

Figure 12.4 shows a scale-free network generated via the copy model with 1,000 vertices. It was grown from a network of just two nodes. Each new node had a 25% chance of adopting each one of an existing node's friends. Now compare this diagram of the copy network with the diagram of the previous network generated through preferential attachment. Even though they are both scale-free networks, they look radically different. The copy network looks a lot more organic. The winding branches look almost like the branches of a tree or the tendrils of a tree's root system.

If you also look closely at the cluster of nodes near the center, you can also see that many of those in the center are clustered. If you had to compare the clustering coefficients of the copy network with that of the preferential attachment network, it would probably be a safe bet that the former would have a higher clustering coefficients. So not only does it have the power-law degree distribution that real-world networks have, but it also has clustering! If you look at the copy process, it is also much more sociologically plausible. Independent of whatever process drives us in choosing friends, once we choose a friend, that friend might introduce us to some of their friends. Thus just like the copy model, we have a chance of becoming friends with a certain number of our new friend's friends.

Alternative Scale-Free Network: The Configuration Model

Friendship networks aren't the only type of social networks out there. Suppose for instance, we wanted to model sexual networks, where one person is connected to the other if they have had sexual relations. Most people probably wouldn't choose a potential sex partner because they have knowledge that the candidate had lots of past partners. In fact, having "been around the block" may negatively impact a candidate's prospects in that department. The copy model wouldn't be a very realistic model either. In terms of sexual networks, it would mean that if a guy has a sexual relationship with a girl, the guy then has a chance with forming sexual relations with a percentage of the girl's previous male partners. There are many reasons why this doesn't happen in the real world, one of which is the fact that "not everyone swings both ways". Sexual networks, however, have a scale-free network distribution. How might we create a scale-free network model that resembles sexual networks without preferential attachment?

One such way is the configuration model (Bender and Canfield, 1978), which can be used to generate any type of random network with a specified distribution, not only a scale-free network. You start with a degree list, or a list of how many connections that each node has in a network. If you want a scale-free network, this degree list would have a power-law distribution. Unlike preferential attachment and the copy model, the configuration model is not a generative model. New nodes are not grown into it; you start with all the nodes that will participate in your network. Another difference with this model is that the power-law degree distribution is something that doesn't emerge from the structure or process of the network; rather, it is something that is predetermined as an attribute of each node.

Let's now examine the process behind the configuration model. While we eventually want it to model networks formed by matches, with two types of nodes (e.g., male and female), for now we will assume only one type of node.

Suppose we have a lot of sheets of paper. On each sheet of paper, we would write the name down of someone in our network. Each paper represents a matchmaking opportunity for the person. Suppose that matchmaking opportunities were distributed like a power law, which would not be too far-fetched if mating market opportunities were based on attributes that were distributed like power laws (e.g., income). So if someone had a lot of matchmaking opportunities, you would write her name down on multiple sheets of paper. Now shuffle the sheets of paper. Scramble them, shake them, hop up and down on one leg, and put them into a hat. Now draw out two sheets of paper. Congratulations, these two lucky names have scored with each other. Record the pair, draw two more names, and repeat until there are no sheets left in the hat and viola, you have a list of connections in your network. Actually, you probably don't need to hop up and down on one leg, but you already knew that.

The larger the network, the less of a chance you have of drawing up two sheets with the same name. This chance approaches zero with very large networks. You have even a lesser chance of drawing up the same two names. And if this ever happens, you could always toss the duplicates back in the hat and give it a good shaking. If you started with a mating-opportunity distribution that was a power law and used a lot of paper, you would end up with a scale-free network.

Figure 12.5 shows what a scale-free network generated by this process would look like. Note that there are significant differences with this network and the networks created from our previous two models. First, although the previous two networks models generated one huge component, this model generates many disconnected pairs, and smaller fragmented components. The majority of nodes, however, are still members of the largest component. Another result is the fact is that the previous scale-free network is a tree is that it is very centralized, whereas in this scale-free network generated from the configuration model, the structure is decentralized. It is harder to imagine a hierarchy or a chain of command. The highly connected hubs are not necessarily at the center, and some nodes with a high degree can exist on the outer rings of the structure.

Although the method we described to create this model did not differentiate between sexes, the structure above is very similar to the structure found in adolescent sexual networks (Bearman et al., 2004). A large proportion of matches exists as pairs. These are our faithful partners. However, there are very few cycles of short lengths. Most cycles are longer and not local. The configuration model could be further tweaked, by modifying it to allow for two types of sexes, and the way that people are matched could be changed. In its most basic form, it provides a good way to generate both a scale-free network and an interesting structure.

In the next section, we will talk about some properties of scale-free networks. However, it is important to keep in mind that scale-free networks come in a variety of styles and that specific properties may vary depending on the brand of scale-free network you are sporting. Again, scale-free networks must all have a power-law degree distribution by definition; however, all the other kind of attributes are up for grabs.

NETWORK DAMAGE AND SCALE-FREE NETWORKS

Suppose an elementary school kid was exploring the properties of ice melting for a school science fair experiment. He wants to know how much an ice cube would melt after being left for an entire day at various temperatures. He has his rich parents buy him a high-tech refrigerator which he can calibrate to exact temperatures, and conducts an experiment.

Each day, he starts off the experiment by setting the refrigerator to a different temperature, say, between −15 and 72 degrees fahrenheit, and observing whether a cube of ice melts. At all the temperatures below freezing, he finds that the ice cube stays in tact. But once he sets the refrigerator to a temperature above the freezing point, the entire ice cube melts. Not half of it, or a part of it; the cube goes from totally frozen to totally melted at a critical point—the freezing point for water. While he had naively expected to find a temperature range where half the ice cube melted and, a few degrees above that, where three-fourths of the ice cube melted, he found that ice melts above the freezing point, but stays solid below. If we think about the effect of temperature on the damage to an ice cube after a day, it is very black and white. This is because the relationship between temperature and the freezing point of water is a threshold, or critical point.

Now suppose we dropped bombs at random locations over a major city. What is the relationship between the amount of bombs dropped and the percentage of buildings destroyed? We would find that that the number of buildings destroyed increases proportionately to the number of bombs we dropped. In fact, we might be able to say something like, for every three bombs we drop, we destroy an average of five buildings. However, thinking back to our ice example, saying that five grams of ice melts for every three degree increase in temperature wouldn't make much sense, would it?

One thing to keep in mind in this chapter is that that many complex and social networks respond to damage the same way that an ice cube responds to increases in temperature. When some types of networks take minor damage, they regenerate and heal in their own way, as traffic, whether in the form of cars, Internet packets, or communication, is directed around the damaged areas. But if the damage to the network is too severe, huge portions, or even the entire network, will be shut down.

With that said, we will first explore how scale-free and random networks are vulnerable to different types of damage. Specifically, we will see how scale-free networks are more vulnerable to calculated damage than are random networks but more resistant to random damage.

Measuring Damage

Damage to networks can be measured using a variety of methods, but most of these methods have to do with the connectivity of the network. For sparsely connected networks, an easy way to define damage is to consider changes based on the composition of components. Recall that a network component is a set of nodes where there exists a path between any two nodes in the set. Holding the number of nodes constant, if the number of components increases, it means something has shattered a vital bridge. Similarly, if the number of components increases with a fixed number of nodes, then it also must mean that the average number of nodes in each component has decreased.

As an example, consider a small town with 100 buildings, with 50 buildings on each side of river. The two halves of the town are connected to each other through a bridge across the river. Suppose one day, high winds destroy the bridge, disconnecting the two halves of the town. In the chaos and confusion, another leader emerges to lead the part of the town without the mayor. When the dust settles, the town decides not to rebuild the bridge and become two separate towns. If you follow the analogy here, the number of components has increased from one to two, and the average size of the components has decreased from 100 to 50.

However, most social networks, and real-world scale-free networks in general, are not as sparsely connected as a road system. We need another way of measuring damage to connectivity without looking at components. Another method of assessing damage is to consider damage to distances among nodes within the network. The simplest way to compute damage would be to simply examine the graph diameter, or the longest geodesic. If the diameter increases, that means distances are being damaged. For example, suppose you wanted to sail from Spain to the Philippines. You would probably use the Panama Canal. Now suppose that the canal was damaged, and ships were unable to pass. You would then probably need to sail around the tip of South America. What was already a long trip was just made longer.

In most cases, even diameter is not sensitive enough to measure the incremental changes caused by damage to a relatively dense network. A very sensitive and widely used method is to simply examine changes to the average shortest distance between all pairs. If no path exists between two nodes, their distance is set to infinity and the harmonic mean can be used, or alternatively one could set their distance to one more than the maximum distance of the network and then calculate the average of the distances. Thus, if the distance between any pair of nodes in the network is increased by disruption (and there are a lot of pairs in a large network), then the characteristic path length would reflect that. Although this measure is very powerful, the downside is that it takes a lot of time to compute for very large networks, as the number of pairs increases dramatically the more nodes there are in the network.

Random vs. Calculated Damage

Now that we have seen how network damage can be measured, we will look at the different ways we can damage networks. Networks can be damaged by removing nodes, or removing the connection between nodes. For example, destroying a road in a transportation network would be similar to removing a connection, while destroying an airport would be similar to destroying a node. In densely connected networks, node removal is likely to be more damaging because the removal of each node is accompanied by the removal of all connections that were associated with that node. In sparsely connected networks the removal of connections themselves is likely to be relatively more damaging because low-density networks are relatively more dependent on their connections to maintain connectivity.

One basic type of damage that networks can be afflicted by is random damage. Sometimes you will hear this type of damage referred to as random failure, because often there is no particular agent inflicting the damage. You may have also heard a network's strength in defending against random failure referred to as robustness. True random damage means that every node or connection in a network has an equal chance of failing. Consider a city road system. Over time, the roads will decay and develop cracks and potholes due to a variety of factors. Such factors include weather, earthquakes, or meteors from outer space. If these were the only factors, damage would be fairly random. However, if cars cause damage to roads (which they do), then it would no longer be considered random failure because roads that are used more by cars would be more likely to receive damage and thus require more maintenance.

The other type of damage we will consider is from calculated attack. A network's strength in defending against calculated attacks is sometimes referred to as resistance or resilience. In this situation, something causes some nodes or connections to be more likely to receive damage. One of the most intuitive and widely used criteria is to simply pick the node with the highest connections. This is equivalent to disabling the airport with the largest number of connecting flights, taking out a website with the greatest amount of links (such as a search engine), or imprisoning the drug cartel member with the most contacts.

Attacking the node with the highest number of connections is the same as attacking the node with the highest degree of centrality. You could probably generalize the criteria for selecting nodes by using other types of centrality. Recall that nodes high in betweeness centrality are nodes that are positioned on a lot of shortest paths between other pairs of nodes. Attacking the node that was most in between other nodes would be equivalent to disabling the airport with the most amount of traffic. So instead of disabling a highly connected airport such as Los Angeles International or New York's John F. Kennedy Airport, you would pick HartsField-Jackson International airport in Atlanta, which is the busiest airport in terms of traffic (but not necessarily connections).

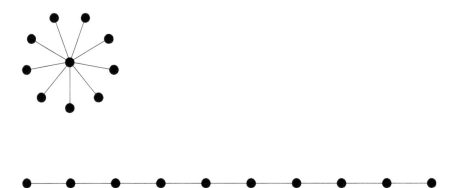

Figure 12.6. Line and star network ($N = 10$).

Another way to vary the criteria for node selection would be to pick proportionately to centrality rather than picking the most central node. If we assumed cars really did substantial damage to roads, and we wanted to simulate possible decay patterns, we would pick roads proportionate to their betweenness centrality to attack. We wouldn't automatically always pick the most traveled road because it's not certain that it would fail first—it is just more likely to fail.

The Effect of Hubs in Scale-Free Networks

The take-home message from this section is that because scale-free networks have so many nodes with so few connections, your average attack wouldn't do that much damage if you picked nodes at random. However, if you knew which node to pick, specifically the hubs, then you would do devastating damage. With random networks, there are always some nodes with more connections than others you could take out, but these nodes would not have a quite the enormous amount of connections that the hubs of scale-free networks would.

Let's illustrate this concept briefly. Figure 12.6 shows two networks, each with 10 nodes and nine connections. The network on the bottom is a line network, and the network on the top is a star network. The line network represents the most egalitarian situation for edge arrangement, with the distribution of degrees as even a possible. Most nodes have two connections, and the nodes on the edges can't help but have one connection. The star network represents the situation with the greatest inequality. The node in the center has nine connections, and all other nodes have only one connection.

Suppose we are measuring damage as an increase in component size and we are attacking at random. In the line network, 8 times out of 10, we will pick one of the nodes with two edges, and disconnect the line into two smaller line segments and damage the network. If we happen to attack one of the two end nodes, we won't disconnect the network, and it will stay in one piece. Under our measure of damage via component size, this is defined as no damage. Now in the star network, 9 times out of 10 we will pick one of the end nodes, resulting in no damage. We only have a 1 in 10 probability of picking the center node, causing catastrophic damage. Thus in the line network, we have an 80% chance of damaging the network, and in the star network, we have a 10% chance of damaging the network.

Now, these networks are small for the sake of illustration. The disparity in damage probability increases as the network grows. If we had a line network with 1,000,000 nodes, we would have a 999,998/1,000,000 chance of damaging the network. In other words, we would almost always inflict damage. With a star node with a million nodes, we have a one in a million chance of inflicting damage. Most of the time, we would do no damage to the network.

However, if we were able to choose which node to attack, we would of course pick the center node in the star network, and cause cataclysmic damage. For the line network, no matter what non end node we pick, we could split the network into only two components.

While they may share similarities, a star network is still far from being a scale-free network; it helps illustrate the mechanic of hubs in scale-free networks because it exaggerates the inequality inherent in a power-law distribution. A line network on the other hand exaggerates the relatively egalitarian nature of a random network.

Cascade Failures

At the beginning of this section, we mentioned that the behavior of scale-free networks to damage may behave more like the threshold response that an ice cube has to temperature. We saw how the structure of scale-free networks causes it to resist random attacks like a suit of armor. Actually a more accurate analogy would be like a lizard shedding off its tail. However, once the severity of an attack reaches a certain point, a network can rapidly shut down, much like a widespread power outage, traffic gridlock, or a coma in the brain. One form of this critical failure is called a cascade failure.

Cascade failures happen in networks where there is something flowing between the connections of the network. The idea is that each connection or node has a certain limit of "traffic" that it can handle. If it becomes overloaded, it will fail as well. When nodes or connections are damaged and removed from the network, but yet the volume of traffic in the network does not decrease, then the traffic must find alternate routes in the network. Nodes and connections that are not used to handling the increased traffic may in turn, fail themselves, causing the traffic to be redirected to other nodes and connections. A chain reaction can soon occur, shutting down the entire network.

Table 12.2.
Distribution of sex partners from the 2008 General Social Survey

Number of sex partners in one year	Cases	Percent
1 or less	1507	85.82
2 partners	120	6.83
3 partners	55	3.13
4 partners	25	1.42
5–10 partners	33	1.88
11–20 partners	11	0.63
21–100 partners	3	0.17
100 + partners	2	0.11
Total	1756	100

Some scale-free networks in the real world are fairly dense. Often even the removal of one hub, while inflicting considerable damage, is not enough to cause critical or cascade failure. However, with the removal of multiple hubs, a scale-free network can rapidly reach this breaking point.

DISEASE SPREAD IN SCALE-FREE NETWORKS

Perhaps one of the most important and interesting properties of scale-free networks is that they are extremely conducive to the spread of disease. Consider a network where people are connected by sexual relations. Table 12.2 contains data on sex partners from the General Social Survey, a popular cross-sectional representative survey of Americans that is administered every few years. Although the attributes to the variables are in categories and not a straight partner count, it is pretty clear that the number of sex partners resembles a power law. The vast majority of people either have 1 or fewer partners, and there were even two people with over 100 different sex partners in a year. That's a lot of partners! If you think about it, they would have to have a new partner almost every three days, or even multiple partners at a time. Or maybe an Amish was sampled, and she thought holding hands was like having sex. Okay, maybe it is better to not think about it. But remember that this is only from a representative sample of 1,756 people.

Sexual networks are fairly hard to map out. It's not as simple as approaching a potential respondent and asking for a list of his current and past sexual partners—especially if he is married and the interviewer is conducting the survey in front of his spouse. But yet, if you are an epidemiologist, or someone who studies the transmission and spread of disease, it is important to know what the network structure looks like in order to forecast how fast a disease can spread. That's where network models come in. Using what we know about the distribution of sex partners, we can generate a number of random scale-free networks and use it in simulations.

What Affects the Spread of Disease?

By definition, a sexually transmitted diseased can be transmitted only through sexual relationships. Of course there are some exceptions such as coming into contact with blood, and so on. But because they must spread through the connections in a network, networks with high distances between nodes or networks with many disconnected components are likely to slow or contain the spread of disease. For example, in a lattice or grid network, the disease would take many steps to fully infect the network. Similarly, in a caveman network with many disconnected clusters, the disease would spread really fast in each cluster, but would not jump to other clusters. Unfortunately for us, distances are low in most network models that we know, such as random, small-world, and scale-free networks, due to random connections that connect what would otherwise be distant parts of the network.

However, scale-free networks make it even easier. In additional to the low distances, scale-free networks also contain hubs, or nodes with a large number of connections. If these nodes become infected, they will infect a great deal many other nodes in the network. Imagine if the two hubs from Table 12.2 were unknowingly infected at the start of the year and did not use condoms—they would have infected over a hundred people.

Many epidemiological models for the spread of disease feature a reproduction number, sometimes called the reproduction rate. This number is usually calculated from the number of newly infected people within the infectious period of the disease from the initial infection in a freshly susceptible population. In random and other types of networks, a disease must have a reproduction rate greater than 1 to become an epidemic. If the disease does not reach the threshold value of the reproduction number, then disease will slowly drop off because not enough new nodes are infected. If the reproduction rate is exactly one, than the disease remains in the population at an constant rate, with the same number of people dying or becoming resistant to the disease as the number of newly infected. In scale-free networks, there is no critical value. Because of the existence of hubs, diseases have the potential to spread independent of the reproduction number. Once a hub node gets sick there is a chance that the disease will pick up and respread.

Although the idea that scale-free networks are one of the best environments that diseases can spread in may seem bad, there are some things researchers and policy makers can use to their advantage. For example, we already know that scale-free networks are exceptionally vulnerable when their hubs are neutralized. Therefore, outreach programs would be both

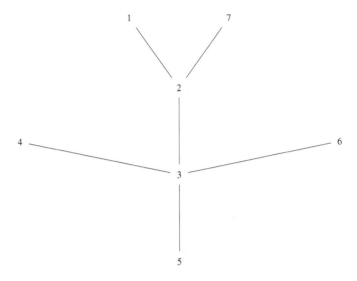

Figure 12.7. Exercise network.

more effective and efficient if they administered education and condoms to people who serve as hubs. Targeted campaigns to at-risk subsets of the population would be vastly more powerful than those trying to reach a broad audience.

EXERCISES

1. Suppose you are given the network in Figure 12.7. Suppose a new node, 8, is entering the network and will choose a new connection based on preferential attachment. For Nodes 1 through 7, calculate their probabilities of forming a connection with the new Node 8.
2. For each of the seven nodes in Figure 12.7, calculate the following caused by its removal from the network: (a) change in the number of components; (b) change in the average path length.

CHAPTER 13

Balance Theory

CLASSIC BALANCE THEORY

You might have heard the old saying: an enemy of my enemy is my friend. If we think of it in social network terms, if B is an enemy of A, and C is an enemy of B, then A and C should be friends. The saying makes intuitive sense, because since A and C are both enemies of B, both of them have a common goal. If they fought each other instead of helped each other, they would only be helping their common enemy, B.

In 1946, social psychologist Fritz Heider took this concept and crafted a theory out of it. He proposed that if two people held positive feelings for each other, then they should be consistent in their opinion toward a third entity. This concept is a bit more general than the words of wisdom that we explored a moment ago. If two people were friends, not only would they share the same enemies, but if one of them were friends with a third person, then they would both be friends with that third person. With Heider's original formulation, you can substitute concepts or ideas as well. For example, if two people are friends, they should both like the same hobbies, or both hate the same kind of foods.

Balanced and Unbalanced Triads

Let's now look at how we can visualize Heider's classic balance theory. Figure 13.1 shows all the possible configurations you can get with three people who either have a positive or negative relation among them. The solid lines represent a positive relationship, and the dotted lines represent a negative relationship. Note that these are also mutual relationships, so the connections are all bidirectional. In this classical representation, every actor must be connected, so there are no empty relationships. You can take a moment to verify that these are indeed all the possible relationships. Each relationship has two possibilities, and there are three relationships, so there are $2 \times 2 \times 2$ or $2^3 = 8$ possible structural configurations.

These eight triads are organized whether or not they are balanced under classical balance theory. The triads enclosed in the box on the left column are balanced, while those on the right are unbalanced. Everyone dislikes

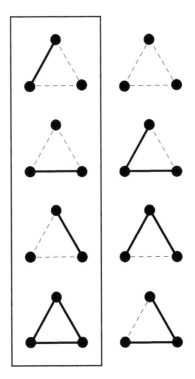

Figure 13.1. Balanced and unbalanced cycles of length 3.

each other in the triad on the top right, which is unbalanced because the enemy of someone's enemy is still his enemy. If this situation described the relationships among three countries, a truce between two countries could happen soon, so that those two countries could stand united against a common enemy.

The three other unbalanced triads in the right column in Figure 13.1 describe a situation where one person has two friends, but the two friends dislike each other. Two things may happen in this case to balance this configuration. The first possibility is the happy ending: the two people who dislike each other decide to kiss and make up, much to the happiness of their shared common friend, and as a result, everyone is friends with each other and there is world peace. The second possibility is more tragic: the antagonism between the two people strains the relationship with their existing friend, until one of the existing positive relationships turns sour, and we are left with one person disliking both of other two.

Formalizing Classical Balance Theory

If we take the assumptions that come with classical balance theory and run with them a bit, we can do interesting stuff. What if we had a

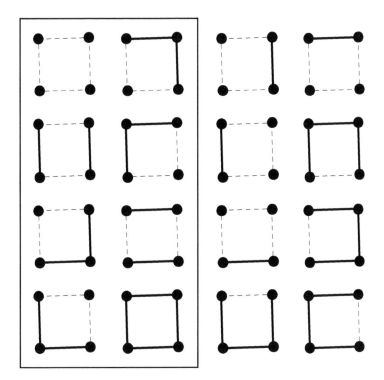

Figure 13.2. Balanced and unbalanced cycles of length 4.

network of positive and negative relationships among a community, and we wanted to assign a score to how "balanced" the network was? We could then take this measure and explore its relationship with other things. For example, suppose a college freshman had data for the social network for each floor of a multistory freshman dormitory. The network representing each floor would be assigned a score based on its social balance. Would this score be correlated to how many drama incidents happen throughout the year on the dorm floor? Does racial diversity in a dorm increase or decrease its social balance? What effect do certain dorm activities led by the residential assistant have on balance? Do certain types of activities increase or decrease the amount of social balance?

To do this, we need a quick and easy way of deciding whether a network is balanced or not. Taking another closer look at Figure 13.1, we may notice something in common with all unbalanced structural configurations: they all have an odd number of negative cycles. Cartwright and Harary (1956) formalized Heider's theory of balance by observing this, and showing that if you multiply all the relationships of a cycle, then if the cycle has an odd number of negative signs, overall the cycle will be negative.

While earlier we looked at cycles with three people, Figures 13.2 and 13.3 show all the possible structural configurations with four and five people, respectively. They're organized in the same way—with balanced

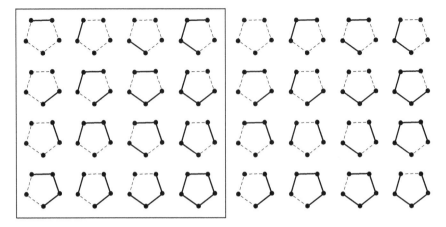

Figure 13.3. Balanced and unbalanced cycles of length 5.

cycles within the box, and unbalanced outside of the box. You can easily see that the unbalanced cycles are the ones with an odd number of negative relationships.

Note that up until this point, the concept of whether something is balanced or not is applied to a cycle, and not a network. A network can contain a lot of cycles. If you're a strict mathematician, an entire network or graph can be balanced if and only if (or "iff" as the mathematicians like to say) every cycle within it is balanced. Whew, that's very strict! As sociologists, we may want a bit more nuanced approach in measuring how balanced a network is. Harary (1959) gave us two suggestions. The first and most simple method is simply to look at all the cycles in a network and ask what proportion of those cycles are balanced. Thus the score would be the number of balanced cycles divided by the total number of cycles.

Another more sophisticated way to measure how balanced a network is would involve counting the number of relationships that must be changed in order for the entire network to be balanced. It could be that one bad cycle is causing a great deal many cycles to be unbalanced, and its removal would fix everything. This can stand in contrast to a situation where you have only a few unbalanced cycles, but each one would require an independent change to balance the situation. The downside to this method is that it may be hard to determine, or at least more computationally intensive to determine the optimal way to balance the network.

Example: A Little Conflict Resolution

Let's explore an example to see how the logic of balance theory can be put into action. Pretend for a moment that there was such thing as a conflict

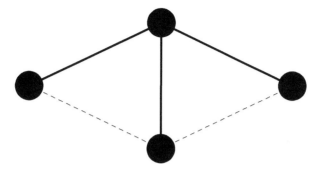

Figure 13.4. The initial conflict situation.

resolution specialist who gets called in to resolve conflicts, and you happen to be one of those specialists. One day, a manager for a small business calls you in for your problem-solving expertise. He tells you that he has three employees working for his company, and that he has positive relationships with all of them. The employees like him, and he likes the employees. The problem is, the employees don't like each other. His first employee doesn't like the second employee, and the second employee doesn't like the third employee. The first and third employee have no relationship with each other because they work at different times.

The situation is causing some social dissonance. It pains him to see his friends hate each other. He wants you end this dissonance because it is causing both him and his employees problems such as loss of sleep at night, awkward group situations at work, and hair loss.

One of the perks of being a special conflict resolution specialist is that you have a magic wand. This magic wand can enigmatically change the social relationship status between two people. But after you use it once, the wand cannot be used anymore on anyone in the same set of people. In a perfect world, you could get everyone to like each other, and then there would be absolutely no conflict because all possible cycles would be balanced. But there are two negative relationships, and you can't rectify them both. In other words, you will need to balance this network by changing one and only one relationship. Which relationship should you change in order to balance the network?

Figure 13.4 shows the situation in graph theoretic form. In this simple network of four people, there are actually three cycles. Two cycles of length three, composed of the manager and two of his employees, and then one big cycle of length four that includes everyone. In the current situation, both of the cycles of length three are unbalanced because they both contain one (an odd number) negative relationship. The cycle of length four is balanced because it contains two negative relationships. Thus, this network is 33% balanced.

Although there are five relationships total in the network, because of network isomorphisms, you really only have three choices. Two pairs of

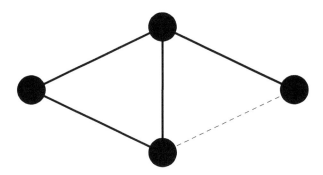

Figure 13.5. Option 1.

choices are equivalent. Let's explore these choices in detail and the impact of each change on the network's overall balance score.

Option 1: Change the relationship between two employees from negative to positive.

You're an optimist, and you want to explore the most positive option first, which is to make love out of hate. If you changed a negative relationship to a positive one, the result would look like Figure 13.5.

Note that if we did not care about the ordering of the people in the network, changing the relationship between the first and middle employee to positive is the same as changing the relationship between the middle and last employee to positive. If we did it this way, one of the unbalanced cycles of length three would become positive. There would be some part of the day where everyone would get along at the workplace because they are all friendly to each other. However, the other cycle of length three remains unbalanced and unchanged.

Furthermore, there is an unexpected side effect from this change. The cycle of length four which was previously balanced because it had two negative relationships is now unbalanced because it has one negative relationship with this change. So we balance a cycle of length three at the cost of unbalancing the cycle of length four. At the end of the day, we still only have one balanced cycle, and the network is still 33 % balanced.

Option 2: Change the relationship between the employer and an outer employee from positive to negative.

Your next option is to change a positive relationship to negative for one of the outer employees. Again, it doesn't matter if you do this for the first or last employee, as you will be left with the same result because of the graph isomorphism.

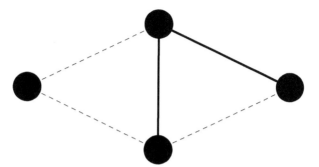

Figure 13.6. Option 2.

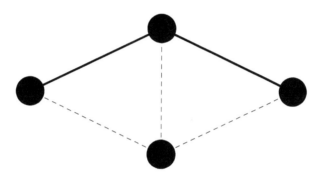

Figure 13.7. Option 3.

If you do it this way, you get the situation represented in Figure 13.6. Much like the first option, you have balanced one of the unbalanced cycles of length three. The difference is you balanced it by getting both the employer and the middle employee to share a common animosity for the outer employee. However, much like the first option, doing this does not change the status of the other cycle of length three, and even unbalanced the cycle of length four. This option leaves the network 33% balanced.

Option 3: Change the relationship between the employer and the middle employee from positive to negative.

Your final option, represented in Figure 13.7, is to turn the positive relationship between the employer and the middle employee from positive to negative. This one looks very promising as you have simultaneously balanced both of the unbalanced cycles of length three. Both of them now contain two negative relationships, which is an even number. At the same time, because the relationship between the employer and the middle employee is not included in the cycle of length four, you leave the balanced cycle unchanged. The network is now 100% balanced because all cycles are balanced.

You are confident in this third option and wave the magical wand. The next day, the middle employee just happens to come to work late, and in casual clothing instead of standard business attire. He also swiped the manager's car in the parking lot on accident and left a nasty scratch. The manager is not pleased, and decides to reduce the middle employee's salary. He also bans the middle employee from parking anywhere near the workplace.

Although a positive relationship was turned negative, there is no more social dissonance and uncomfortable situations at work. Everyone is clear on how they should feel—they hate the middle guy and the middle guy hates them. During breaks the manager stands by the water cooler with either of the outer employees and gossips about the misfortunes of the middle employee. Your mission is complete.

The Meaning of Cycle Length

In the example above, we gave the same weight to cycles in our calculation, regardless of their length. For example, both Option 1 and Option 2 had a end result of unbalancing the cycle of length four, while turning a unbalanced cycle of length three into a balanced one. We concluded this was not a net gain because we lost a balanced cycle in order to gain one.

However, in reality, this probably did help a bit. This is because the exact social meaning of cycles beyond lengths of three is a bit more complicated. Cycles beyond lengths of three don't represent immediate social dissonance among their members. One possible idea they could represent is future opportunities for dissonance. This idea is based on the assumption that people will meet each other through transitivity. Classical measures of graph balance acknowledged the different meaning of larger cycles by reducing the weight of their contribution to overall graph balance. So, theoretically, you could argue that Options 1 and 2 presented a small improvement in overall balance of the social configuration.

Because of this cloudy area for large cycles in classic balance theory, future directions focused exclusively on the triad as the unit of analysis, and disregarded larger cycles. We will see this in more detail when we discuss structural balance.

The Micro-Macro Link

Sociologists are often concerned with the micro-macro link, that is, how the behavior of a set of actors operating based on their localized social environments can lead to emergence of large-scale patterns. One reason why balance theory was popularized among researchers was that it had a strong macro-level implication that was the logical result of the micro-level assumptions being true. This is sometimes referred to as the structural theorem of balance theory.

The world according to balance theory is a place where you can divide everyone into two sets of people. Everyone in each set has either a positive or no relationship with people in the same set, while having a negative or no relationship with people in the other set. The key point here is that if there was a balanced world, people didn't "choose" which side that they wanted to belong to on the large scale; they simply acted on the local level by preferring that their friends shared the same views on other friends as they do. The large-scale organization is a result of the system of local actions.

Whether the world can truly be thought of as moving toward this two faction system remains an open question. However it does help explain why it is hard to remain neutral in large-scale political configurations, such as the Allies and Axis powers during World War II. The modern-day analog would be the industrialized world versus what President George W. Bush controversially termed the "Axis of Evil." Even within party politics, the right or left tries to win over the moderate independent voters in a close election.

Another reason why the world powers or even your own set of friends might not easily be divided into two easily discernable factions is that there are constantly factors that change relationships outside of balance theory. Balance theory is ultimately a theory of structural endogenous change. You are given a set of people and relationships, but other than the dots and lines in your graph, you really don't know anything else about the people. You don't know if one person is ugly or handsome, rich or poor, outgoing or antisocial, or any other attributes or preferences they may have. All you know is the structure of relations. In the real world, other things can cause social change that is not a result of structure itself. These are often referred to as exogenous factors.

STRUCTURAL BALANCE

Perhaps the most immediate spiritual successor to classical balance theory is structural balance. This line of research started to not just theorize but explore options for testing some ideas within balance theory empirically. Structural balance has some notable differences from classical balance theory.

First, the unit of analysis was no longer a cycle of relationships, but rather triads. So no matter what the configuration of a network, the number of triads examined will always be based on the number of people in the network. With cycles, there can be none in a network, or there can be even more cycles than triads.

Secondly, while balance theory looked at two types of relationships, structural balance examines only a single dichotomous relationship. This is significant because the structure of many more types of networks can be analyzed. Classical balance theory fits only relationships that satisfy the criteria of antithetical duality, which meant that there was a positive and negative relationship, both mutually exclusive, and two negative relationships would amount to a positive relationship.

Third, structural balance considers the direction of relationships while classical balance was bidirectional. A lot of social relationships may not necessarily be mutual. For example, a person can respect another without receiving respect in return.

The Triad Census

Given the format of structural difference discussed above, there is a consistent set of possible structural configurations. For any given triad, a relationship can either be mutual, directed one way, directed in reverse, or null. This yields $4 \times 4 \times 4$ or $4^3 = 64$ possible structural configurations. However, as you may have noticed with the 8 triad configurations in classical balance, much of these raw triads are simply rotations or reflections of each other.

Thus researchers have come up with a consistent system of nomenclature to describe the unique structural configurations used in structural balance. It turns out that the 64 possible structural configurations given by triplets can be reduced to 16 classes of triads. These triads were given a name based on the number of mutual, asymmetric, or null relationships they contain, respectively. Therefore, the system is called MAN labeling.

The 16 classes of MAN triads and the triplets they represent are given in Table 13.1.

These 16 MAN triads are often seen as the building blocks of social networks, in the same way DNA serves as the building blocks for genetic code or elements are building blocks for the periodic table. By decomposing a network into a triad census of MAN triads, a lot can be gleaned about what is actually going on in the network. Much like classical balance, structural balance theorists argued that certain types of MAN triads were inherently more instable than others. The criteria to determine whether a triad was balanced or not were dependent on the specific theory.

For example, clustering theories were the first generalization of social balance. Recall that social balance allowed for a world where there were essentially two factions. Clustering relaxes this, allowing for more than just two factions. However, there some other conditions that the triads have to follow in order to be "unbalanced" or "balanced."

For example, according to hierarchal clustering, an asymmetrical relationship represents a relationship between people belonging to two clusters on different levels. So if A likes B and B doesn't like A in return, then B belongs to a higher "social" cluster than A. If A likes B and B likes A in

Table 13.1.
Sixty-four triplets classified into 16 MAN triads

MAN label	Triplets	# of triplets
003		1
012		6
102		3
021D		3
021U		3
021C		6
111D		6
111U		6
030T		6
030C		2
201		3
120D		3
120U		3
120C		6
210		6
300		1

Figure 13.8. Balance exercise.

return, they are in the same social cluster, and if neither A or B likes each other at all, then they are in different social clusters, but on the same social level.

So every triad presented in Table 13.1 can be seen as violating this rule or conforming to it. For example, Triad 300 is considered balanced because all the people are in the same cluster with mutual relationships. However, Triad 210 violates this rule and is unbalanced. One person has mutual friendships with two people, implying that all three people are in the same social cluster. However, one of the person's friends likes the other, and the other doesn't like them back, implying that one of his friend is on a higher social circle than the other. Both of these situations can't be true at the same time, so this contradiction makes Triad 210 "unbalanced."

Davis (1970) explored the evidence for structural balance by examining almost 800 social networks. He found that for the most part, the "balanced" triads according to the clustering theorems appeared at higher rates than the "unbalanced" triads, lending support for the new version of balance theory. However, there were some triads that were unbalanced that still seemed to occur often (210 was one of them). More social theories came out to explain this, and balance theory further evolved.

EXERCISES

1. For the following sequence of edges, assumed to be in a cycle, determine whether the cycle is balanced or not.

   (a) $+ + + - + + -$
   (b) $- - -$
   (c) $+ + + + - -$
   (d) $+ + - +$

2. For each cycle in the Figure 13.8, determine whether the cycle is balanced or not.
3. For each of the 16 triad types in Table 13.1, determine whether it is permitted under the rules of structural balance.

CHAPTER 14

Markov Chains

This chapter will cover a probabilistic method for describing how individuals or systems change over time. The technique is Markov chains, named after a 19th century Russian mathematician. Potentially, it can be used to describe phenomena as diverse as fluctuations in the stock market, intergenerational social mobility, and gambling losses. Please review Chapter 3 on probability before continuing.

EXAMPLES

Markov Chains Example 1

Markov chains describe situations in which a system moves among various states in a simple probabilistic way that is different from statistical independence. Suppose there is a bowl with two red and two blue balls and that you sample from the bowl with replacement every time you withdraw a ball. The events are independent, and the probability of drawing six blue balls in a row is $(1/2)^6$. Now suppose you sample without replacement. The probability of drawing a blue ball varies with the present composition of the bowl, and you can't draw six consecutive blue balls. In a Markov chain the probability of moving from one state to another depends on the current state only. Markov chains are described by transition matrices, which give the probabilities that a system moves among its various states. The rows have been labeled with the numbers of red and blue balls describing that state of the bowl. Note drawing blue balls in a pair of neighboring trials are not independent events: having drawn a blue ball in one event reduces the probability of drawing a blue ball in future events.

$$
\begin{array}{c}
2R2B \\
2R1B \\
2R0B \\
1R2B \\
1R1B \\
1R0B \\
0R2B \\
0R1B \\
0R0B
\end{array}
\begin{pmatrix}
0 & 1/2 & 0 & 0 & 1/2 & 0 & 0 & 0 & 0 \\
0 & 0 & 1/3 & 0 & 2/3 & 0 & 0 & 0 & 0 \\
0 & 0 & 0 & 0 & 0 & 1 & 0 & 0 & 0 \\
0 & 0 & 0 & 0 & 2/3 & 0 & 1/3 & 0 & 0 \\
0 & 0 & 0 & 0 & 0 & 1/2 & 0 & 1/2 & 0 \\
0 & 0 & 0 & 0 & 0 & 0 & 0 & 0 & 1 \\
0 & 0 & 0 & 0 & 0 & 0 & 0 & 1 & 0 \\
0 & 0 & 0 & 0 & 0 & 0 & 0 & 0 & 1 \\
0 & 0 & 0 & 0 & 0 & 0 & 0 & 0 & 1
\end{pmatrix}
\qquad (14.1)
$$

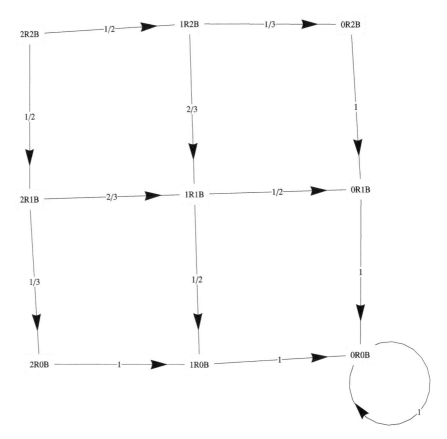

Figure 14.1. Removing balls from a bowl.

A Markov process can also be represented by a graph in which the vertices are states and the edge weights are transition probabilities. Figure 14.1 shows a graph for this example.

Example: Movement among Political Parties

Suppose that in a community in which all voters are Republican or Democrat, the probabilities that Republicans and Democrats change parties between elections are .05 and .10, respectively. There are two states, Republican and Democrat. Let the two rows of the transition matrix be the party, Republican or Democrat, that someone votes for in one election, and let the columns refer to the party that the person votes for in the next election. The transition matrix shows the probabilities that someone who votes for the row party in one election will vote for the column party in the

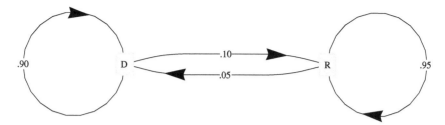

Figure 14.2. A graph representation for party switching.

next election. The transition matrix **P** is as follows:

$$\mathbf{P} = \begin{pmatrix} .95 & .05 \\ .1 & .9 \end{pmatrix} \tag{14.2}$$

Figure 14.2 shows the graph of this two-state Markov chain.

The existence of a transition matrix between states does not mean that we have a Markov process. Movement between political parties is almost certainly not a Markov process. In a Markov process the probability of a transition depends only on the current state, and this is not true for transitions among parties. In each class there will be new flighty voters and independents who switch easily between parties and those with a firm party identification who never change. Such heterogeneity means that the probability of a move does not depend solely on the current state, and the implications we can draw using the Markov assumption will not be true, as we shall show later.

Example 3—Switching Marbles in Two Bowls

Suppose that there are two bowls, each bowl containing four marbles, and that there are four blue and four red marbles. We randomly choose one marble from each bowl and transfer them.

The system is clearly entirely characterized by the number of blue marbles in the left bowl. If there are x blue marbles in the left bowl, then there must be 4-x red marbles in the left bowl, 4-x blue marbles in the right bowl, and x red marbles in the right bowl. Therefore, the rows and columns refer to the number of blue marbles in the left bowl. The rows refer to the number of blue marbles in the left bowl before a switch, and the columns to the number of blue marbles in the left bowl after a switch.

The transition matrix among the five states (there are zero, one, two, three, or four blue marbles in the left bowl) is given by the following transition matrix P. If there are no blue marbles in the left bowl, a switch of one marble from each bowl must result in one blue marble in the left

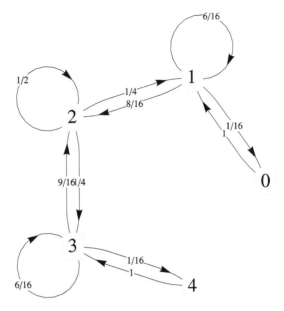

Figure 14.3. Switching marbles in two bowls.

bowl. The calculation of the actual values in P is beyond this course, but note the pattern.

$$P = \begin{pmatrix} 0 & 1 & 0 & 0 & 0 \\ \frac{1}{16} & \frac{6}{16} & \frac{9}{16} & 0 & 0 \\ 0 & \frac{1}{4} & \frac{1}{2} & \frac{1}{4} & 0 \\ 0 & 0 & \frac{9}{16} & \frac{6}{16} & \frac{1}{16} \\ 0 & 0 & 0 & 1 & 0 \end{pmatrix} \tag{14.3}$$

Figure 14.3 shows the graph for this transition matrix.

Example 4—The Rat in a Maze

Suppose there is a rat in a maze composed of nine rooms arranged on a three by three grid. Each room has a opening to the adjacent room. The nine states will be the room the rat is in. We assume that the rat is equally likely to take any of the openings going from one room to another. The following transition matrix shows the probabilities that the rat will move from one room to another after a change of room. The rows are the room the rat is currently in. The columns are where the rat is after one move.

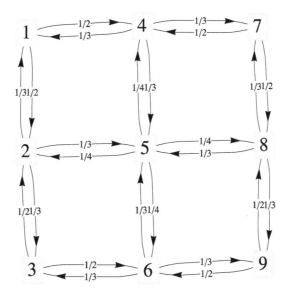

Figure 14.4. Graph of rat maze.

The elements of the matrix are the probabilities of a given change.

$$\mathbf{P} = \begin{pmatrix} 0 & \frac{1}{2} & 0 & \frac{1}{2} & 0 & 0 & 0 & 0 & 0 \\ \frac{1}{3} & 0 & \frac{1}{3} & 0 & \frac{1}{3} & 0 & 0 & 0 & 0 \\ 0 & \frac{1}{2} & 0 & 0 & 0 & \frac{1}{2} & 0 & 0 & 0 \\ \frac{1}{3} & 0 & 0 & 0 & \frac{1}{3} & 0 & \frac{1}{3} & 0 & 0 \\ 0 & \frac{1}{4} & 0 & \frac{1}{4} & 0 & \frac{1}{4} & 0 & \frac{1}{4} & 0 \\ 0 & 0 & \frac{1}{3} & 0 & \frac{1}{3} & 0 & 0 & 0 & \frac{1}{3} \\ 0 & 0 & 0 & \frac{1}{2} & 0 & 0 & 0 & \frac{1}{2} & 0 \\ 0 & 0 & 0 & 0 & \frac{1}{3} & 0 & \frac{1}{3} & 0 & \frac{1}{3} \\ 0 & 0 & 0 & 0 & 0 & \frac{1}{2} & 0 & \frac{1}{2} & 0 \end{pmatrix} \qquad (14.4)$$

Figure 14.4 is the graph of this transition matrix.

Gambling

In the "matching pennies" game, two people both hold a penny in their hand. They simultaneously reveal whether the coins are heads or tails. If the coins have the same face showing (both heads or both tails), one of the players wins both pennies. If they have different faces showing, then the other player wins and he gets both pennies. Suppose that two people A and B are equally adept at matching pennies, and there are five pennies between them. The system has six states corresponding to the number of

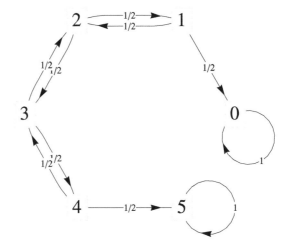

Figure 14.5. Graph for the penny game.

pennies possessed by Player A. Suppose, moreover, that play stops when all pennies are possessed by one of the players. The transition matrix would look as follows:

$$\mathbf{P} = \begin{pmatrix} 1 & 0 & 0 & 0 & 0 & 0 \\ \frac{1}{2} & 0 & \frac{1}{2} & 0 & 0 & 0 \\ 0 & \frac{1}{2} & 0 & \frac{1}{2} & 0 & 0 \\ 0 & 0 & \frac{1}{2} & 0 & \frac{1}{2} & 0 \\ 0 & 0 & 0 & \frac{1}{2} & 0 & \frac{1}{2} \\ 0 & 0 & 0 & 0 & 0 & 1 \end{pmatrix} \qquad (14.5)$$

Figure 14.5 shows the graph.

States 0 and 5 are called absorbing states; once entered, they cannot be left. Such states may be identified by a "1" on the main diagonal of the transition matrix.

POWERS OF P, PATHS IN THE GRAPHS, AND LONGER INTERVALS

In the chapter on matrix multiplication we learned that powers of the adjacency matrix gave us the numbers of walks of various lengths between vertices. Specifically $(\mathbf{A}^k)_{ij}$ is the number of walks of length k from vertex i to vertex j. Similarly, walks in these diagrams of transition matrices of independent events correspond to sequences of states through which the system passes. The product of the probabilities on a path corresponds to the probability that the system will pass through that sequence of states.

Let's look at $(\mathbf{P}^2)_{ij}$.

$$(\mathbf{P}^2)_{ij} = p_{i1}p_{1j} + p_{i2}p_{2j} + \ldots + p_{in}p_{nj} \tag{14.6}$$

Each product gives the probability that the system will pass from state i to an intermediate state and then from the intermediate state to j because of the Markov property that probabilities of transitions depend on the current state of the system. Since the sum is over all intermediate states, $(\mathbf{P}^2)_{ij}$ gives the probability that the system will be in state j two moves later if it starts in state i.

$(\mathbf{P}^k)_{ij}$ **equals the probability that the system will be in state** j **after** k **moves if it starts in state** i.

For example, suppose the mouse starts out in Room 1. The first row of $\mathbf{P}^k$ gives the probabilities that the mouse will be in various of the other rooms after k moves. The second row of $\mathbf{P}^k$ gives similar information if the mouse starts in the second row. The following matrices show the probabilities of mouse destinations after 5 and 10 changes of room.

$$\mathbf{P}^5 = \begin{pmatrix} 0 & 0.28 & 0 & 0.28 & 0 & 0.22 & 0 & 0.22 & 0 \\ 0.19 & 0 & 0.19 & 0 & 0.33 & 0 & 0.15 & 0 & 0.15 \\ 0 & 0.28 & 0 & 0.22 & 0 & 0.28 & 0 & 0.22 & 0 \\ 0.19 & 0 & 0.15 & 0 & 0.33 & 0 & 0.19 & 0 & 0.15 \\ 0 & 0.25 & 0 & 0.25 & 0 & 0.25 & 0 & 0.25 & 0 \\ 0.15 & 0 & 0.19 & 0 & 0.33 & 0 & 0.15 & 0 & 0.19 \\ 0 & 0.22 & 0 & 0.28 & 0 & 0.22 & 0 & 0.28 & 0 \\ 0.15 & 0 & 0.15 & 0 & 0.33 & 0 & 0.19 & 0 & 0.19 \\ 0 & 0.22 & 0 & 0.22 & 0 & 0.28 & 0 & 0.28 & 0 \end{pmatrix} \tag{14.7}$$

$$\mathbf{P}^{10} = \begin{pmatrix} 0.17 & 0 & 0.17 & 0 & 0.33 & 0 & 0.17 & 0 & 0.16 \\ 0 & 0.25 & 0 & 0.25 & 0 & 0.25 & 0 & 0.25 & 0 \\ 0.17 & 0 & 0.17 & 0 & 0.33 & 0 & 0.16 & 0 & 0.17 \\ 0 & 0.25 & 0 & 0.25 & 0 & 0.25 & 0 & 0.25 & 0 \\ 0.17 & 0 & 0.17 & 0 & 0.33 & 0 & 0.17 & 0 & 0.17 \\ 0 & 0.25 & 0 & 0.25 & 0 & 0.25 & 0 & 0.25 & 0 \\ 0.17 & 0 & 0.16 & 0 & 0.33 & 0 & 0.17 & 0 & 0.17 \\ 0 & 0.25 & 0 & 0.25 & 0 & 0.25 & 0 & 0.25 & 0 \\ 0.16 & 0 & 0.17 & 0 & 0.33 & 0 & 0.17 & 0 & 0.17 \end{pmatrix} \tag{14.8}$$

W can see that there is a sense in which no equilibrium remerges over time: the zeros in the transition matrix change their position radically. That's because the mouse always moves from an odd-numbered to an even-numbered room. Let's look again at the system of Example 1, where there were Republican and Democratic voters. The following shows the transition matrices after 5 and 20 time periods.

$$\mathbf{P}^5 = \begin{pmatrix} 0.81 & 0.19 \\ 0.37 & 0.63 \end{pmatrix} \tag{14.9}$$

$$\mathbf{P}^{20} = \begin{pmatrix} 0.68 & 0.32 \\ 0.64 & 0.36 \end{pmatrix} \tag{14.10}$$

The first row of $\mathbf{P}^k$ gives the probabilities that someone who starts Republican will be a Republican or a Democrat k elections later. The second row has a similar interpretation for those who begin as Democrats. Here the probabilities seem to be approaching stability. Moreover, it appears that the long-range probability that any voter is Republican is about two thirds, regardless of her initial party identification.

THE MARKOV ASSUMPTION: HISTORY DOES NOT MATTER

It was stated that P^k gives the probabilities after k transitions. A closer examination of the equation for P^2 shows that this is true only if a particular assumption is true.

$$(P^2)_{ij} = p_{i1}p_{1j} + p_{i2}p_{2j} + \ldots + p_{in}p_{nj} \tag{14.11}$$

$p_{i1}p_{1j}$ is supposed to represent the probability that the system will move from state i to state 1 and then on to state j. But, we can use the multiplication rule only if the events are independent. The probability of a transition from state i to state j must be unaffected by whether the system had been in state i previously or in some other state. This is not always true.

For example, suppose that there are two kinds of voters, those who never change parties and those who flit from party to party according to the whim of the moment. Call the first type party *stalwarts* and the second *independents*. Current Republicans will consist of both stalwarts and independents. The probability of a transition to the Democratic Party will be zero for stalwarts and something greater than that for independents. If a Republican had previously been a Democrat, we can infer that he must be an independent, and therefore his probability of changing to the Democrats would be greater than if he had regularly voted Republican and is therefore more likely to be a stalwart.

In this case, the probability of moving from Republican to Republican is not independent of past moves, and the above equation would not be valid for inferring the probabilities of sequences of transitions. This is always a question we must ask in any application of the Markov chain model.

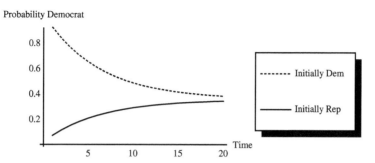

Figure 14.6. Converging probabilities for Example 2.

TRANSITION MATRICES AND EQUILIBRIUM

Let's look at Example 2 again, where **P** described transitions between the Republican and Democratic Parties. P^k gives the probabilities of the two states after k moves. Let's graph the probabilities that those who are initially Republican or Democrat are Republican or Democrat after k elections.

The two probabilities become indistinguishable after enough time. You can see this also by looking at high powers of P, as in Equation 14.10. In this last matrix, the rows are all almost equal. This common row gives the probabilities of the various states. This common vector of long-run probabilities is called an *equilibrium vector*. Some Markov matrices have equilibrium vectors of long-term probabilities that are independent of the starting state.

Not all do, however. We have seen already that in the rat-maze example, after an even number of moves the rat would be in the same kind of room he started out in (odd or even), and after an odd number of moves he would have to be in a different kind of room. Therefore, there is no equilibrium. If, for example, the rat starts in Room 1, after 20 moves he would have a .167 chance of being in Room 1, but after 21 moves he would have no chance of being in Room 1. The probability of being in Room 1 oscillates between .167 and zero, never reaching equilibrium.

Consider now the gambling example. There is no single equilibrium that describes this process. The long-run probabilities depend on the starting state, as given by the following matrix.

$$\mathbf{P}^{\infty} = \begin{pmatrix} 1. & 0 & 0 & 0 & 0 & 0 \\ 0.8 & 0 & 0 & 0 & 0 & 0.2 \\ 0.6 & 0 & 0 & 0 & 0 & 0.4 \\ 0.4 & 0 & 0 & 0 & 0 & 0.6 \\ 0.2 & 0 & 0 & 0 & 0 & 0.8 \\ 0 & 0 & 0 & 0 & 0 & 1. \end{pmatrix} \tag{14.12}$$

There are, in fact, three types of equilibrium behavior in transition matrices. One type never arrives at equilibrium for some or all starting states because of some kind of periodicity (the maze Example 4). Others arrive at a set of equilibrium vectors that depend on the starting state (the gambling Example 5). Others arrive at an equilibrium set of probabilities that do not depend on the initial probabilities (Examples 1, 2, and 3). There are mathematical criteria for determining each type of transition matrix (Hoel et al., 1972, Chapter 2). We will determine the type by examining high powers of the transition matrix.

For example, let's take Example 3.

$$\mathbf{P}^\infty = \begin{pmatrix} 0.014 & 0.229 & 0.514 & 0.229 & 0.014 \\ 0.014 & 0.229 & 0.514 & 0.229 & 0.014 \\ 0.014 & 0.229 & 0.514 & 0.229 & 0.014 \\ 0.014 & 0.229 & 0.514 & 0.229 & 0.014 \\ 0.014 & 0.229 & 0.514 & 0.229 & 0.014 \end{pmatrix} \quad (14.13)$$

Every row gives the same long-term probabilities of being in each state regardless of the initial state.

Chapter Demonstrations

- Markov Chains

EXERCISES

1. There are two parties in a country, the Social Democrats (S) and the Christian Democrats (C). The following matrix shows the probabilities of voter movement between the parties, with the first row and column being C.

$$\mathbf{P} = \begin{pmatrix} 0.7 & 0.3 \\ 0.1 & 0.9 \end{pmatrix}$$

   (a) Treating movement as a Markov chain, what is the probability that a Social Democrat is a Social Democrat for the next two elections?
   (b) What is the probability that a Social Democrat will be a Social Democrat again two elections later?
   (c) What is the long-term distribution of party membership?

2. Which of the following matrices are transition matrices? Where not, list the assumptions that are violated. Of those that are truly transition matrices, which will settle down to a long-term equilibrium set of

probabilities independent of the initial state?

$$\mathbf{A} = \begin{pmatrix} 0 & 0 & 1 \\ 0 & 1 & 0 \\ 1 & 0 & 1 \end{pmatrix} \quad \mathbf{B} = \begin{pmatrix} .5 & .5 & .0 \\ .3 & .4 & .4 \\ .1 & .1 & .8 \end{pmatrix}$$

$$\mathbf{C} = \begin{pmatrix} .1 & .6 & .3 \\ .1 & .6 & .3 \\ .1 & .6 & .3 \end{pmatrix} \quad \mathbf{D} = \begin{pmatrix} .6 & .0 & .4 \\ .0 & 1.0 & .0 \\ .7 & .0 & .3 \end{pmatrix}$$

$$\mathbf{E} = \begin{pmatrix} 0.7 & 0.3 \\ 0.3 & 0.7 \end{pmatrix} \quad \mathbf{F} = \begin{pmatrix} 1 & 0 & 0 \\ 0 & 1 & 0 \\ 0 & 0 & 1 \end{pmatrix}$$

3. The following transition matrix describes the hypothetical probabilities that a child of a member of the row class (upper, middle, lower) is destined to be a member of the column social class.

$$\mathbf{P} = \begin{pmatrix} .7 & .3 & 0 \\ .1 & .6 & .3 \\ .1 & .1 & .8 \end{pmatrix}$$

   (a) If social mobility were a Markov process, what would be the probability that the grandchild of a member of the upper class would also be upper class?
   (b) What would be the probability that the grandchild of a member of the lower class would also be power class?
   (c) From this matrix, can you determine the probability that the grandfather of an upper-class individual was also upper class? Why, or why not?
   (d) If in one generation 25% of the population is upper class, 30% is middle class, and 45% is lower class, what will be the distribution of classes in the next generation?
   (e) What is the long-range equilibrium distribution of class size?

4. Think of reasons why social mobility between social classes would not be a Markov process. Does the fact that wealthy parents can afford more education for their children lead to a violation? What about inherited wealth?

5. Two cross-town football rivals play football every year. They agree that, starting from the present, if one team wins the annual game two years in a row the trophy associated with the rivalry will be retired to the campus of the winning ream. If the probability that Team A wins is p and the probability that Team B wins is $q = 1 - p$, create the five-state transition matrix for this contest.

6. Using the demonstration Markov Chains, describe both situations and the Markov chains that model them with the following properties. Do not use examples from the chapter.

(a) There is no equilibrium but rather an oscillation between two states.
(b) The equilibrium distribution depends on the initial conditions.
(c) The system inevitably ends up in a single absorbing state.
(d) There is a single equilibrium distribution independent of the initial state.

CHAPTER 15
Demography

Demography is about counting populations and studying population processes. The value of being able to accurately count people cannot be understated. Imagine if a chemist tried to study the properties of molecules without any weighing scale to determine mass. Imagine if a construction worker tried to build a skyscraper without a tool resembling a ruler. A serious scientifically minded sociologist would always have a tool from the tool box of demography in hand.

In addition to being a method, demography itself is an important substantive area of study. Being able to forecast where a population is headed and understand why is invaluable. In some ways, the consequences of population change are similar to those of climate change. The incremental changes behind most population processes are hard to detect in the short term. However, if catastrophic changes occur over the long term, it may already be too late to take any action on it.

For example, take Japan, a country that is experiencing one of the most underappreciated problems of the 21st century: low fertility and rapid aging. In just a few decades from now, half of Japan's population will be over 60. Imagine if one out of every two faces you see on the street is an elderly person, almost unable to walk independently. Several decades from now, Japan's population will be cut in half. Imagine the economic consequences as a shrinking working-age population has to support a population too old to work, with the proportion of youth declining each year due to low birth rates. Infrastructure such as schools and homes becomes unused, and the costs of health care skyrocket due to the care of the elderly. If nothing is done, populations can face extinction.

The irony is that classical demographic theory warned us of the opposite scenario. The late-18th-century political economist Thomas Malthus, considered by many to be the fountainhead of demography, cautioned that the world's population would exponentially explode and outstrip technological advances in food production and land use. Resources would be used up and life would be miserable. His theoretical influence reached into the 20th century, where world leaders implemented polices that mandated one child per household or subsidized and provided incentives for women to tie off their fallopian tubes.

Figure 15.1. Life and death as a process.

Through modern demography, the consensus is that such a scenario is unlikely. Birth rates naturally fall after a drop in mortality rates, in what demographers call the first demographic transition. The theory is that all societies that undergo industrialization will eventually transition to a society where people survive longer and give birth to fewer children. Even countries with high fertility rates such as India and Mexico will have their growth rates flatten out.

Demography is a rich area of practical research in the social sciences, and a proper introduction would require an entire course, or even a sequence of courses. However, using the framework introduced in the preceding chapters of this book, we can examine the core process of demography, population projection, through the lens of a mathematical sociologist.

MORTALITY

Death is the easiest aspect of a population process to model, so let's explore it first.

Example: Life and Death

Figure 15.1 shows a simple diagram of the process of life and death. It can be thought of as a network with two vertices, each vertex representing the state of life and death, respectively. The relations indicate flow from one state to another. An individual who is alive can stay alive, or she can die and move into the death state. An individual who is dead stays dead and, unless we believe in zombies or reincarnation, will never be in the state of life ever again (even zombies are not truly alive, and deserve their own state: undead).

$$\begin{pmatrix} 1 & 1 \\ 0 & 1 \end{pmatrix} \tag{15.1}$$

Equation 15.1 gives the adjacency matrix representing the diagram. Note that every value is filled in except for the value that symbolizes movement from death to life. We can also turn this into a Markov process, which we covered in the previous chapter, by setting probabilities for life and death so that the value of each row is equal to 1. This is done in Equation 15.2. Because the death entry contains a value of 1, death meets the definition for an "absorbing state" in our Markov process.

$$\begin{pmatrix} .9 & .1 \\ 0 & 1 \end{pmatrix} \tag{15.2}$$

Projecting an initial population of 1,000 people in the alive state can be accomplished through matrix multiplication, as we saw in the last chapter. Equations 15.3 through 15.5 project our initial population three time frames into the future, where we have 729 people alive and 271 dead.

$$\begin{pmatrix} 1000 & 0 \end{pmatrix} \cdot \begin{pmatrix} .9 & .1 \\ 0 & 1 \end{pmatrix} = \begin{pmatrix} 900 & 100 \end{pmatrix} \tag{15.3}$$

$$\begin{pmatrix} 900 & 100 \end{pmatrix} \cdot \begin{pmatrix} .9 & .1 \\ 0 & 1 \end{pmatrix} = \begin{pmatrix} 810 & 190 \end{pmatrix} \tag{15.4}$$

$$\begin{pmatrix} 810 & 190 \end{pmatrix} \cdot \begin{pmatrix} .9 & .1 \\ 0 & 1 \end{pmatrix} = \begin{pmatrix} 729 & 271 \end{pmatrix} \tag{15.5}$$

Eventually, everyone will die, and our population vector will become $\begin{pmatrix} 0 & 1000 \end{pmatrix}$. Death is always the one great certainty in this world.

Example: Life Cycle of Cats

In the previous example with life and death, the relations represented the flow of people. Individuals flowed from life to death, and they were allowed to remain within same state. In our next example, there will be three major changes. First, we are going to reverse the meaning of the relations. Here $a \rightarrow b$ if the population at state a "draws upon" the population at state b. The reason why we do this is to bring our transition matrix closer to the structure of an actual population projection matrix. The second difference is that life will not simply be one state but multiple states. Individuals are not allowed to stay in one state. As time passes, they grow older and enter states that represent older ages. Finally, we remove the death state. It will be assumed that if an individual does not advance to the next age state, that the individual has met his unfortunate destiny and passed on.

Figure 15.2 gives the network diagram that shows the structure of our new population process with the changes described above. Here, the individuals in the population live at most 9 years. Obviously, this probably doesn't represent a human population, since we tend to live a little bit longer than 9 years. So let's make this example represent the life cycles of cats. Domesticated cats can live longer than 9 years of course; some grow as old as 20. But just play along with me here; it doesn't have to be purrfect.

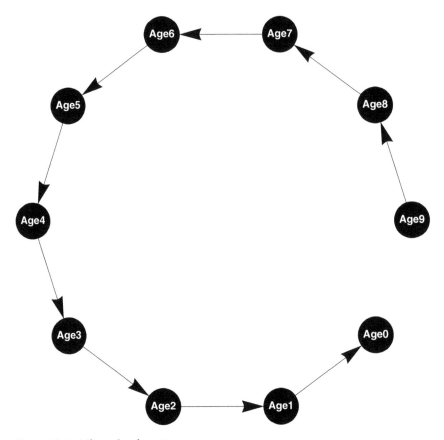

Figure 15.2. Life cycle of a cat.

Again, relations mean the same thing as in our first example, Age0 → Age1, because cats move from being 0 years of age to 1 year of age. But in Figure 15.2, relations represent states drawing upon the state that the arrow is pointed toward. So Age1 → Age0, because the population of cats of 1 year of age is drawn from the population of cats at 0 years of age.

$$
\begin{pmatrix}
0 & 0 & 0 & 0 & 0 & 0 \\
{}_np_0 & 0 & 0 & 0 & 0 & 0 \\
0 & {}_np_{1n} & 0 & 0 & 0 & 0 \\
0 & 0 & {}_np_{2n} & 0 & 0 & 0 \\
0 & 0 & 0 & \ldots & 0 & 0 \\
0 & 0 & 0 & 0 & {}_np_{(N-1)n} & 0
\end{pmatrix}
\tag{15.6}
$$

The general form of the transformation matrix representing such a change will resemble something that looks like Equation 15.6. Since we're getting a bit more formal, $_np_x$ represents the probability that an individual who starts off at age x will live the following n years. For example, $_5p_{20}$ would represent the probability that an individual who is 20 years old would survive 5 more years to the age of 25. This lowercase p is used pretty consistently by demographers. Along the same lines, $_nq_x$ represents the probability that an individual who starts off at age x will die sometime in the following n years. In general n is the period length of the state, and x is the period number. In our example with cats, $n = 1$ and x has a maximum of 9. Technically there are 10 states ($N = 10$) because we include $x = 0$ as a state. The question of how to calculate these probabilities based on empirical data is outside the scope of an introductory chapter on demography, from the angle of a mathematical sociologist. In this chapter we will just learn how to work with these numbers.

$$_np_x = 1 - {_nq_x} \tag{15.7}$$

The probability of dying and the probability of survival have a simple relationship described by Equation 15.7. They both have to add up to 1. So any time you have the probability of death, you can deduce the probability of survival, and vice versa.

In the previous example with life and death, the population vector was horizontal; now the population vector is vertical. We will denote the population vector at time t as P_t. Note that this is uppercase P, not lowercase. Let's generate the hypothetical probabilities for feline survival. These probabilities, along with our initial population of 100 cats, are represented in Equation 15.8.

$$A = \begin{pmatrix} 0 & 0 & 0 & 0 & 0 & 0 & 0 & 0 & 0 & 0 \\ .95 & 0 & 0 & 0 & 0 & 0 & 0 & 0 & 0 & 0 \\ 0 & .9 & 0 & 0 & 0 & 0 & 0 & 0 & 0 & 0 \\ 0 & 0 & .85 & 0 & 0 & 0 & 0 & 0 & 0 & 0 \\ 0 & 0 & 0 & .8 & 0 & 0 & 0 & 0 & 0 & 0 \\ 0 & 0 & 0 & 0 & .75 & 0 & 0 & 0 & 0 & 0 \\ 0 & 0 & 0 & 0 & 0 & .7 & 0 & 0 & 0 & 0 \\ 0 & 0 & 0 & 0 & 0 & 0 & .65 & 0 & 0 & 0 \\ 0 & 0 & 0 & 0 & 0 & 0 & 0 & .6 & 0 & 0 \\ 0 & 0 & 0 & 0 & 0 & 0 & 0 & 0 & .55 & 0 \end{pmatrix}, \quad P_o = \begin{pmatrix} 100 \\ 0 \\ 0 \\ 0 \\ 0 \\ 0 \\ 0 \\ 0 \\ 0 \\ 0 \end{pmatrix} \tag{15.8}$$

In general, the population vector in the next time frame is equal to the transition matrix multiplied by the population vector in the current time frame, or $P_{t+1} = A.P_t$. Note that because our population vector is vertical instead of horizontal, we place the transition matrix on the left side so we don't violate matrix multiplication rules.

$$P_1 = A.P_0 = \begin{pmatrix} 0 & 0 & 0 & 0 & 0 & 0 & 0 & 0 & 0 & 0 \\ .95 & 0 & 0 & 0 & 0 & 0 & 0 & 0 & 0 & 0 \\ 0 & .9 & 0 & 0 & 0 & 0 & 0 & 0 & 0 & 0 \\ 0 & 0 & .85 & 0 & 0 & 0 & 0 & 0 & 0 & 0 \\ 0 & 0 & 0 & .8 & 0 & 0 & 0 & 0 & 0 & 0 \\ 0 & 0 & 0 & 0 & .75 & 0 & 0 & 0 & 0 & 0 \\ 0 & 0 & 0 & 0 & 0 & .7 & 0 & 0 & 0 & 0 \\ 0 & 0 & 0 & 0 & 0 & 0 & .65 & 0 & 0 & 0 \\ 0 & 0 & 0 & 0 & 0 & 0 & 0 & .6 & 0 & 0 \\ 0 & 0 & 0 & 0 & 0 & 0 & 0 & 0 & .55 & 0 \end{pmatrix} \cdot \begin{pmatrix} 100 \\ 0 \\ 0 \\ 0 \\ 0 \\ 0 \\ 0 \\ 0 \\ 0 \\ 0 \end{pmatrix} = \begin{pmatrix} 0 \\ 95 \\ 0 \\ 0 \\ 0 \\ 0 \\ 0 \\ 0 \\ 0 \\ 0 \end{pmatrix}$$

(15.9)

$$P_2 = A.P_1 = \begin{pmatrix} 0 & 0 & 0 & 0 & 0 & 0 & 0 & 0 & 0 & 0 \\ .95 & 0 & 0 & 0 & 0 & 0 & 0 & 0 & 0 & 0 \\ 0 & .9 & 0 & 0 & 0 & 0 & 0 & 0 & 0 & 0 \\ 0 & 0 & .85 & 0 & 0 & 0 & 0 & 0 & 0 & 0 \\ 0 & 0 & 0 & .8 & 0 & 0 & 0 & 0 & 0 & 0 \\ 0 & 0 & 0 & 0 & .75 & 0 & 0 & 0 & 0 & 0 \\ 0 & 0 & 0 & 0 & 0 & .7 & 0 & 0 & 0 & 0 \\ 0 & 0 & 0 & 0 & 0 & 0 & .65 & 0 & 0 & 0 \\ 0 & 0 & 0 & 0 & 0 & 0 & 0 & .6 & 0 & 0 \\ 0 & 0 & 0 & 0 & 0 & 0 & 0 & 0 & .55 & 0 \end{pmatrix} \cdot \begin{pmatrix} 0 \\ 95 \\ 0 \\ 0 \\ 0 \\ 0 \\ 0 \\ 0 \\ 0 \\ 0 \end{pmatrix} = \begin{pmatrix} 0 \\ 0 \\ 85.5 \\ 0 \\ 0 \\ 0 \\ 0 \\ 0 \\ 0 \\ 0 \end{pmatrix}$$

(15.10)

Equations 15.9 and 15.10 show the computational setup needed to obtain the number of cats during the two time frames following our initial time frames. Out of our initial population of 100 cats, 5 die because $_1p_0 = 0.95$, and we are left with 95 cats at the second time frame. The fact that these cats are now 1 year old instead of 0 years old is represented by the number being represented as the second element in the vector rather than the first. For the third time frame, our population of 95 one-year-old cats becomes 85.5 two-year-old cats, computed from 95×0.9. Can you figure out how many four-year-old cats there will be next year?

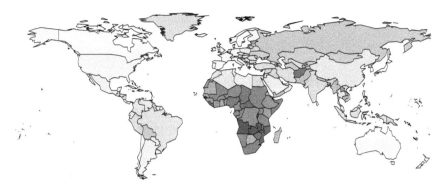

Figure 15.3. World countries shaded by life expectancy.

LIFE EXPECTANCY

Figure 15.3 shows a map of the world, where countries are shaded based on their life expectancy of the population. If you had to take a guess, would you say that the darker colors represent countries with shorter life expectancies or longer life expectancies?

Life expectancy is the quintessential demographic statistic. It is a quick measure of a population's general health and living conditions. Deadly wars, violence, infectious disease, starvation, and poor health care are just some of the factors that can put a dent in your life expectancy. The sign of an industrialized nation is often based on the strength of its life expectancy.

In *The Lord of the Rings: The Two Towers*, King Theoden of Rohan, while burying his slain son, remarks, "No parent should have to bury their child." However, throughout most of the history of mankind, life expectancy has been extremely low. The predominant reason behind this was a high childhood mortality rate. More children died at birth and were susceptible to diseases at young ages. Thus in the health and technological conditions of a medieval-themed society, the scene of parents burying their children was actually fairly common (Get over it Mr. Theoden!). To keep a population growing or, at the very least, to maintain a population, people had to produce many children in the hopes that some would survive.

The turning point, or what demographers refer to as the first demographic transition, is heralded by a turning point in child mortality. Before this point, a parent was more likely to witness the death of her child than the child witnessing the death of his parent. After a population passes this transition point, it is more likely for a child to outlive his parents. This single factor has had the most impact on the life expectancy statistic. We will see why as we explore how this statistic is computed.

Life Expectancy Is Not Average Age at Death

Although most people no doubt have heard the term "life expectancy," very few of those who do not study demography or social sciences will know what exactly the term represents. The most common erroneous answer is that the life expectancy of a country at a specific year is the average age of those people who died in that year. It is worth exploring why demographers don't use this simple average as a measure of health conditions for a country.

Suppose we have a population where only one person died in the current year. Say, for example, that this person was 60 years old. Can we say that the life expectancy of this population is 60? First we need to know the history of this person, to be called John Doe from this point on. Let's say John Doe was an immigrant and moved to this population when he was 30 years old. So for half of his life, John Doe was being exposed to the health conditions of another population that is not the current population. If the society he emigrated from had horrible health conditions, then he probably died young relative to what we would expect in the current population.

Even if John Doe had not immigrated but lived in his current population for all of his life, his age of death is still not a good indicator of life expectancy. The mortality conditions in this society might have changed during his life course. For example, say that water treatment methods were invented and implemented when John Doe was 30. Antibiotics were invented when John Doe was 40. Then for most of John Doe's life, these two things that would no doubt improve the health conditions of his society didn't exist. So the young children who are growing up in this population will no doubt have a healthier future than John Doe.

In general, the age at which a person dies is determined by a history that spans across space and time. It may or may not be indicative of health conditions of the society in which that person is living at the time of death. While the actual measure of life expectancy is far from perfect, it is still a better measure of the current health conditions in a society than simply considering age at death.

Definition

Life expectancy, e_x, is not a simple number, it is a function where you input an age. Thus, demographers often speak of life expectancy "at age x," which is the number of years you would be expected to live if you survived to age x in your current population. The most common form of the life expectancy is e_0, which is also called life expectancy at birth. Although child mortality is much lower in industrialized societies, it is still higher than usual for the very first year, which may bias the life expectancy statistic, as we will later see. Because life expectancy represents the additional years you are expected to live at age x, the age that you are expected to die is presented by $x + e_x$.

To calculate e_x, or the life expectancy at age x, you add up all of the age-specific survival probabilities at x or greater and multiply by the period length. This concept is embodied in Equation 15.11.

$$e_x = \sum_{i=x}^{N-1}({}_n p_{in})n \tag{15.11}$$

It is worth noting that life expectancy is also measured on different levels of precision. The form you are given above is a very crude form. A more exact form would take into account the time contribution of those who died, by multiplying half of the time in each period with the number of people who died or $(1 - {}_n p_{in})\frac{n}{2}$. Those who are even more exact will estimate the area under a person-years-lived curve. But let's keep it simple and just worry about how life expectancy is defined in Equation 15.11.

Example: Life Expectancy of Our Cats

$$A = \begin{pmatrix} 0 & 0 & 0 & 0 & 0 & 0 & 0 & 0 & 0 & 0 \\ .95 & 0 & 0 & 0 & 0 & 0 & 0 & 0 & 0 & 0 \\ 0 & .9 & 0 & 0 & 0 & 0 & 0 & 0 & 0 & 0 \\ 0 & 0 & .85 & 0 & 0 & 0 & 0 & 0 & 0 & 0 \\ 0 & 0 & 0 & .8 & 0 & 0 & 0 & 0 & 0 & 0 \\ 0 & 0 & 0 & 0 & .75 & 0 & 0 & 0 & 0 & 0 \\ 0 & 0 & 0 & 0 & 0 & .7 & 0 & 0 & 0 & 0 \\ 0 & 0 & 0 & 0 & 0 & 0 & .65 & 0 & 0 & 0 \\ 0 & 0 & 0 & 0 & 0 & 0 & 0 & .6 & 0 & 0 \\ 0 & 0 & 0 & 0 & 0 & 0 & 0 & 0 & .55 & 0 \end{pmatrix}$$

Recall the transition matrix that represents the probability of survival for cats, which I reproduced above so you don't have to flip back a few pages (I'm a nice guy). What is the life expectancy at birth, or e_0, for our cats?

$$e_0 = \sum_{i=0}^{9}({}_1 p_{i(1)})1 = (.95 + .9 + .85 + .8 + .75 + .7 + .65 + .6 + .55)(1) = 6.75 \tag{15.12}$$

Thus, given A, our cats are expected to live 6.75 years from the age that they are born. Because each state in our transition matrix represents 1 year, $n = 1$. Next question, given that a cat lives to be 3 years of age,

(a) what is the life expectancy of that cat, e_3, and (b) what is the expected age of death? Equation 15.13 shows that the cat is expected to live 4.05 more years, and Equation 15.14 shows that if we add these 4.05 years to the cat's current age of 3 years, the cat is expected to die at an age of 7.05 years.

$$e_3 = \sum_{i=3}^{9} (_1 p_{i(1)})1 = (.8 + .75 + .7 + .65 + .6 + .55)(1) = 4.05 \qquad (15.13)$$

$$x + e_x = 3 + 4.05 = 7.05 \qquad (15.14)$$

One interesting thing to take note of here is that this older cat is expected to die at an older age than the cat who was just born. This is because we know that this cat has already survived to the age of three. So in the first three years, this cat beat the odds and did not die from being exposed to the .95, .9, and .85 survival probabilities of the first three years. So it makes sense that this cat will live a bit longer than a cat who was just born and whose fate in the early years of life is uncertain.

Example: Life Expectancy of a Harsh Human Society

$$A = \begin{pmatrix} 0 & 0 & 0 & 0 & 0 & 0 & 0 & 0 \\ .90 & 0 & 0 & 0 & 0 & 0 & 0 & 0 \\ 0 & .99 & 0 & 0 & 0 & 0 & 0 & 0 \\ 0 & 0 & .95 & 0 & 0 & 0 & 0 & 0 \\ 0 & 0 & 0 & .9 & 0 & 0 & 0 & 0 \\ 0 & 0 & 0 & 0 & .8 & 0 & 0 & 0 \\ 0 & 0 & 0 & 0 & 0 & .6 & 0 & 0 \\ 0 & 0 & 0 & 0 & 0 & 0 & .3 & 0 \end{pmatrix} \qquad (15.15)$$

Equation 15.15 gives the transition matrix of death for a harsh human society. In contrast to our cat example, each state now represents a period of 10 years ($n = 10$). So .90 represents the probability of being born at age 0 and surviving to age 10. With 8 states, this matrix tracks the probability of survival up until the age 80 bracket. Equation 15.16 gives the life expectancy at birth for this society, which works out to be 54.4 years.

$$e_0 = \sum_{i=0}^{7} (_{10} p_{i(10)})10 = (.90 + .99 + .95 + .9 + .8 + .6 + .3)(10) = 5.44(10) = 54.4$$

$$(15.16)$$

The Meaning of Life (Expectancy)

So what exactly does life expectancy measure? If you think about it, it's a measure that is not representative of any individual in a population. As we noted at the beginning of this section, individuals carry with them their own unique history of exposure to health conditions. The real interpretation of life expectancy at birth for a society is that if we froze the health conditions of a particular society at that specific time frame, and then took a hypothetical person and had that person live her entire life in that time frame, life expectancy would represent the expected age of death for this hypothetical person.

An alternative way of calculating life expectancy is through a life table. It is essentially a social simulation, but through a spreadsheet, where an initial population of sims, sometimes called the radix, is exposed to all the age-specific probabilities of survival. The person's years lived contributed by the total simulation of sims is then summed, and then divided by the initial population. This type of spreadsheet simulation is one of the major themes of demographic methods, along with decomposition. There are more nuanced, complex, and profound variations of this style, but just from life expectancy you can get a grasp of the underlying idea.

Thus life expectancy is not representative of any individual person in the society, but merely a measure of health at that snapshot in time. It best represents the present, but not the past nor the future.

FERTILITY

Death is just one part of a larger population process that includes birth and immigration. In this section we will further evolve our transition matrix that encapsulates the process of death to include the process of new life, and we will arrive at what demographers call the Leslie matrix, or the matrix that can be used to project forward populations in time. It is named after Paul H. Leslie, who first came up with the method in 1945.

Again, let's examine the network diagram represented by the process of mortality to visualize the structure of births. This graph is given in Figure 15.4. Remember that the relations represent "draws from." So we see that the population of cats at age 0 is drawn upon by cats aged 2–4. We are entertaining the idea, which may or may not be true, that 1-year-old cats are too young to have kittens, and that cats that are 5 years old or older are too old to have kittens.

So what number will go into our Leslie matrix to represent births? While the numbers that represent deaths in our matrix are probabilities, the numbers that represent births are rates. It is possible for a person to have multiple births within a period of time, but not possible to have multiple

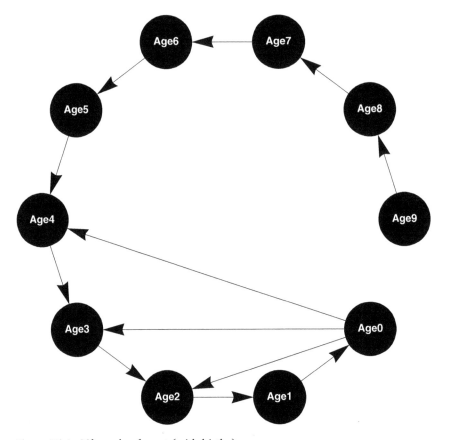

Figure 15.4. Life cycle of a cat (with births).

deaths—although you know what they say about cats and having nine lives! The distinction between probability and rate is very, very important in demography. This is also one of those things that those unattuned to this craft will use interchangeably. It's not easy for the average person to tell you the difference between a death rate and a probability of death.

In the chapter on probability earlier in this book, we learned that probability was the fraction of an event over the sum of all events. In demography, probability is the fraction of the number of occurrences over the number of preceding events or trials, whereas rate is the fraction of the number of occurrences over the number of person-years lived. For example, the probability of divorce is the number of divorces over its preceding event, marriage. The probability of death is the number of deaths overs its preceding event, life. The divorce rate is the number of divorces over the combined time of everyone who is married. The death rate is the number of deaths in a population over the combined time of those spent alive.

Thus the number that will go into our Leslie matrix for births is called the age-specific maternity rate, or formally, $_nF_x$. So, for example, $_5F_{20}$ represents the number of babies born to women between the ages of 20 and 25, divided by the number of person-years lived by women between 20 and 25. To further complicate things, this figure assumes a sex ratio of 0.5, or a population with the same number of females as males. Because males can't reproduce, we really care only about females, and babies who are born females. So sometimes a one-sex model is used and the symbol that represents the age-specific maternity rate becomes $_nF_x^F$. It can get really complicated when you start considering parity issues, birth intervals, and so on, but for now let's accept the fact that the number that will end up in our Leslie matrix will represent the number of babies born to those in that state.

For example, let's suppose that the age-specific maternity rate for cats at age 2 will be 0.75, $_1F_2 = 0.75$. This means that if we have 1,000 cats living in that age range, they will produce $1,000 \times 0.75 = 750$ kittens. Not only are these kittens are cute, but they will no doubt be added to the population aged 0 years for the next time frame. Let's assume that $_1F_3 = 1.25$ and $_1F_4 = 0.75$; Equation 15.17 shows what our Leslie matrix would look like.

$$
\begin{pmatrix}
0 & 0 & .75 & 1.25 & .75 & 0 & 0 & 0 & 0 & 0 \\
.95 & 0 & 0 & 0 & 0 & 0 & 0 & 0 & 0 & 0 \\
0 & .9 & 0 & 0 & 0 & 0 & 0 & 0 & 0 & 0 \\
0 & 0 & .85 & 0 & 0 & 0 & 0 & 0 & 0 & 0 \\
0 & 0 & 0 & .8 & 0 & 0 & 0 & 0 & 0 & 0 \\
0 & 0 & 0 & 0 & .75 & 0 & 0 & 0 & 0 & 0 \\
0 & 0 & 0 & 0 & 0 & .7 & 0 & 0 & 0 & 0 \\
0 & 0 & 0 & 0 & 0 & 0 & .65 & 0 & 0 & 0 \\
0 & 0 & 0 & 0 & 0 & 0 & 0 & .6 & 0 & 0 \\
0 & 0 & 0 & 0 & 0 & 0 & 0 & 0 & .55 & 0
\end{pmatrix}
\tag{15.17}
$$

POPULATION PROJECTION

We'll now step through a population projection step-by-step with all that we covered in this chapter, using our example with cats. Remember that population projection is based on the strong assumption that mortality and fertility will stay constant into future time frames, an assumption that may or may not necessarily hold. In the worst case scenario, it gives us a counterfactual to use as a standard of comparison. It shows us "what would be," "if we assume this."

$$\begin{pmatrix} 0 & 0 & .75 & 1.25 & .75 & 0 & 0 & 0 & 0 & 0 \\ .95 & 0 & 0 & 0 & 0 & 0 & 0 & 0 & 0 & 0 \\ 0 & .9 & 0 & 0 & 0 & 0 & 0 & 0 & 0 & 0 \\ 0 & 0 & .85 & 0 & 0 & 0 & 0 & 0 & 0 & 0 \\ 0 & 0 & 0 & .8 & 0 & 0 & 0 & 0 & 0 & 0 \\ 0 & 0 & 0 & 0 & .75 & 0 & 0 & 0 & 0 & 0 \\ 0 & 0 & 0 & 0 & 0 & .7 & 0 & 0 & 0 & 0 \\ 0 & 0 & 0 & 0 & 0 & 0 & .65 & 0 & 0 & 0 \\ 0 & 0 & 0 & 0 & 0 & 0 & 0 & .6 & 0 & 0 \\ 0 & 0 & 0 & 0 & 0 & 0 & 0 & 0 & .55 & 0 \end{pmatrix} \cdot \begin{pmatrix} 1000 \\ 0 \\ 0 \\ 0 \\ 0 \\ 0 \\ 0 \\ 0 \\ 0 \\ 0 \end{pmatrix} = \begin{pmatrix} 0 \\ 950. \\ 0 \\ 0 \\ 0 \\ 0 \\ 0 \\ 0 \\ 0 \\ 0 \end{pmatrix}$$

(15.18)

We will start with an initial population of 1,000 kittens. Equation 15.18 shows the matrix multiplication setup necessary to age these kittens. Based on the popular Internet meme and its variations, God killed 50 kittens in our first time frame, leaving 950 one-year-old kittens. Our newborn kittens are certainly too young to understand where kittens come from, so they make no kittens of themselves. Thus in the next time frame, there are no kittens aged 0 years.

$$\begin{pmatrix} 0 & 0 & .75 & 1.25 & .75 & 0 & 0 & 0 & 0 & 0 \\ .95 & 0 & 0 & 0 & 0 & 0 & 0 & 0 & 0 & 0 \\ 0 & .9 & 0 & 0 & 0 & 0 & 0 & 0 & 0 & 0 \\ 0 & 0 & .85 & 0 & 0 & 0 & 0 & 0 & 0 & 0 \\ 0 & 0 & 0 & .8 & 0 & 0 & 0 & 0 & 0 & 0 \\ 0 & 0 & 0 & 0 & .75 & 0 & 0 & 0 & 0 & 0 \\ 0 & 0 & 0 & 0 & 0 & .7 & 0 & 0 & 0 & 0 \\ 0 & 0 & 0 & 0 & 0 & 0 & .65 & 0 & 0 & 0 \\ 0 & 0 & 0 & 0 & 0 & 0 & 0 & .6 & 0 & 0 \\ 0 & 0 & 0 & 0 & 0 & 0 & 0 & 0 & .55 & 0 \end{pmatrix} \cdot \begin{pmatrix} 0 \\ 950. \\ 0 \\ 0 \\ 0 \\ 0 \\ 0 \\ 0 \\ 0 \\ 0 \end{pmatrix} = \begin{pmatrix} 0 \\ 0 \\ 855. \\ 0 \\ 0 \\ 0 \\ 0 \\ 0 \\ 0 \\ 0 \end{pmatrix}$$

(15.19)

Equation 15.19 shows the population at the next time frame, in which more kittens die and we are sadly left with fewer kittens than we started out with. So far this process looks identical to the earlier mortality-only process that we examined, with the only difference being that we started out with 10 times more kittens. This is because we haven't reached the stages where these cats are old enough to have kittens themselves.

$$
\begin{pmatrix}
0 & 0 & .75 & 1.25 & .75 & 0 & 0 & 0 & 0 & 0 \\
.95 & 0 & 0 & 0 & 0 & 0 & 0 & 0 & 0 & 0 \\
0 & .9 & 0 & 0 & 0 & 0 & 0 & 0 & 0 & 0 \\
0 & 0 & .85 & 0 & 0 & 0 & 0 & 0 & 0 & 0 \\
0 & 0 & 0 & .8 & 0 & 0 & 0 & 0 & 0 & 0 \\
0 & 0 & 0 & 0 & .75 & 0 & 0 & 0 & 0 & 0 \\
0 & 0 & 0 & 0 & 0 & .7 & 0 & 0 & 0 & 0 \\
0 & 0 & 0 & 0 & 0 & 0 & .65 & 0 & 0 & 0 \\
0 & 0 & 0 & 0 & 0 & 0 & 0 & .6 & 0 & 0 \\
0 & 0 & 0 & 0 & 0 & 0 & 0 & 0 & .55 & 0
\end{pmatrix}
\cdot
\begin{pmatrix}
0 \\ 0 \\ 855. \\ 0 \\ 0 \\ 0 \\ 0 \\ 0 \\ 0 \\ 0
\end{pmatrix}
=
\begin{pmatrix}
641.25 \\ 0. \\ 0. \\ 726.75 \\ 0. \\ 0. \\ 0. \\ 0. \\ 0. \\ 0.
\end{pmatrix}
$$

$$(15.20)$$

The setup for the third time frame is shown with Equation 15.20. A miracle has happened here. The 855 cats that started off the period being 2 years of age were old enough to have kittens, and they produced according to their age-specific maternity rate, which was 0.75. So we now have $855 \times 0.75 = 641.25$ new kittens in our population. This also gave a significant boost to our total cat population, which jumped up from 855 to $641.25 + 726.75 = 1{,}368$.

$$
\begin{pmatrix}
0 & 0 & .75 & 1.25 & .75 & 0 & 0 & 0 & 0 & 0 \\
.95 & 0 & 0 & 0 & 0 & 0 & 0 & 0 & 0 & 0 \\
0 & .9 & 0 & 0 & 0 & 0 & 0 & 0 & 0 & 0 \\
0 & 0 & .85 & 0 & 0 & 0 & 0 & 0 & 0 & 0 \\
0 & 0 & 0 & .8 & 0 & 0 & 0 & 0 & 0 & 0 \\
0 & 0 & 0 & 0 & .75 & 0 & 0 & 0 & 0 & 0 \\
0 & 0 & 0 & 0 & 0 & .7 & 0 & 0 & 0 & 0 \\
0 & 0 & 0 & 0 & 0 & 0 & .65 & 0 & 0 & 0 \\
0 & 0 & 0 & 0 & 0 & 0 & 0 & .6 & 0 & 0 \\
0 & 0 & 0 & 0 & 0 & 0 & 0 & 0 & .55 & 0
\end{pmatrix}
\cdot
\begin{pmatrix}
641.25 \\ 0. \\ 0. \\ 726.75 \\ 0. \\ 0. \\ 0. \\ 0. \\ 0. \\ 0.
\end{pmatrix}
=
\begin{pmatrix}
908.438 \\ 609.188 \\ 0. \\ 0. \\ 581.4 \\ 0. \\ 0. \\ 0. \\ 0. \\ 0.
\end{pmatrix}
$$

$$(15.21)$$

$$
\begin{pmatrix}
0 & 0 & .75 & 1.25 & .75 & 0 & 0 & 0 & 0 & 0 \\
.95 & 0 & 0 & 0 & 0 & 0 & 0 & 0 & 0 & 0 \\
0 & .9 & 0 & 0 & 0 & 0 & 0 & 0 & 0 & 0 \\
0 & 0 & .85 & 0 & 0 & 0 & 0 & 0 & 0 & 0 \\
0 & 0 & 0 & .8 & 0 & 0 & 0 & 0 & 0 & 0 \\
0 & 0 & 0 & 0 & .75 & 0 & 0 & 0 & 0 & 0 \\
0 & 0 & 0 & 0 & 0 & .7 & 0 & 0 & 0 & 0 \\
0 & 0 & 0 & 0 & 0 & 0 & .65 & 0 & 0 & 0 \\
0 & 0 & 0 & 0 & 0 & 0 & 0 & .6 & 0 & 0 \\
0 & 0 & 0 & 0 & 0 & 0 & 0 & 0 & .55 & 0
\end{pmatrix}
\cdot
\begin{pmatrix}
908.438 \\ 609.188 \\ 0. \\ 0. \\ 581.4 \\ 0. \\ 0. \\ 0. \\ 0. \\ 0.
\end{pmatrix}
=
\begin{pmatrix}
436.05 \\ 863.016 \\ 548.269 \\ 0. \\ 0. \\ 436.05 \\ 0. \\ 0. \\ 0. \\ 0.
\end{pmatrix}
$$

$$(15.22)$$

$$
\begin{pmatrix}
0 & 0 & .75 & 1.25 & .75 & 0 & 0 & 0 & 0 & 0 \\
.95 & 0 & 0 & 0 & 0 & 0 & 0 & 0 & 0 & 0 \\
0 & .9 & 0 & 0 & 0 & 0 & 0 & 0 & 0 & 0 \\
0 & 0 & .85 & 0 & 0 & 0 & 0 & 0 & 0 & 0 \\
0 & 0 & 0 & .8 & 0 & 0 & 0 & 0 & 0 & 0 \\
0 & 0 & 0 & 0 & .75 & 0 & 0 & 0 & 0 & 0 \\
0 & 0 & 0 & 0 & 0 & .7 & 0 & 0 & 0 & 0 \\
0 & 0 & 0 & 0 & 0 & 0 & .65 & 0 & 0 & 0 \\
0 & 0 & 0 & 0 & 0 & 0 & 0 & .6 & 0 & 0 \\
0 & 0 & 0 & 0 & 0 & 0 & 0 & 0 & .55 & 0
\end{pmatrix}
\cdot
\begin{pmatrix}
436.05 \\ 863.016 \\ 548.269 \\ 0. \\ 0. \\ 436.05 \\ 0. \\ 0. \\ 0. \\ 0.
\end{pmatrix}
=
\begin{pmatrix}
411.202 \\ 414.247 \\ 776.714 \\ 466.028 \\ 0. \\ 0. \\ 305.235 \\ 0. \\ 0. \\ 0.
\end{pmatrix}
$$

$$(15.23)$$

Equations 15.21 through 15.23 shows the population distribution for the fourth, fifth, and sixth time frames. The total sums of the population vector are 2,099.03, 2,283.38, and 2,373.4, respectively. Notice that this time span represents a period of slow growth, as our initial population of cats, of which only 305.235 remain, grew too old to have kittens. Growth in the sixth time frame is maintained only by the first wave of the second generation of cats. This generation, which was 548.269 strong, produced 411.202 kittens.

$$
\begin{pmatrix}
0 & 0 & .75 & 1.25 & .75 & 0 & 0 & 0 & 0 & 0 \\
.95 & 0 & 0 & 0 & 0 & 0 & 0 & 0 & 0 & 0 \\
0 & .9 & 0 & 0 & 0 & 0 & 0 & 0 & 0 & 0 \\
0 & 0 & .85 & 0 & 0 & 0 & 0 & 0 & 0 & 0 \\
0 & 0 & 0 & .8 & 0 & 0 & 0 & 0 & 0 & 0 \\
0 & 0 & 0 & 0 & .75 & 0 & 0 & 0 & 0 & 0 \\
0 & 0 & 0 & 0 & 0 & .7 & 0 & 0 & 0 & 0 \\
0 & 0 & 0 & 0 & 0 & 0 & .65 & 0 & 0 & 0 \\
0 & 0 & 0 & 0 & 0 & 0 & 0 & .6 & 0 & 0 \\
0 & 0 & 0 & 0 & 0 & 0 & 0 & 0 & .55 & 0
\end{pmatrix}
\cdot
\begin{pmatrix}
411.202 \\ 414.247 \\ 776.714 \\ 466.028 \\ 0. \\ 0. \\ 305.235 \\ 0. \\ 0. \\ 0.
\end{pmatrix}
=
\begin{pmatrix}
1165.07 \\ 390.641 \\ 372.823 \\ 660.207 \\ 372.823 \\ 0. \\ 0. \\ 198.403 \\ 0. \\ 0.
\end{pmatrix}
$$

$$(15.24)$$

Equation 15.24 represents the seventh time frame, and another baby boom in our cat population. This is because for the first time in our cat population's history, there are two age groups simultaneously contributing to the number of new kittens. The number of new births skyrockets to 1,165.07 in the next year, and the total population of cats jumps to 3,159.97.

Matrix Powers

Recall that for social networks, the power of the adjacency matrix represented the number of paths that exist with length equal to the power.

For example, the square of an adjacency matrix representing friendship gives an adjacency matrix where people are connected if they share a common friend.

Multiplying transition matrices has a similar meaning for population projection. Instead of doing the projection step-by-step, we can simply take the power of the Leslie matrix and multiply it directly to the initial population vector to jump to the population distribution represented by that matrix power. For example, suppose we take the seventh power of the Leslie matrix, or $A.A.A.A.A.A.A$, given by Equation 15.25.

$A^7 =$

$$
\begin{pmatrix}
1.16507 & 1.45736 & 1.23501 & 0.775632 & 0.308401 & 0. & 0. & 0. & 0. & 0. \\
0.390641 & 1.16507 & 1.53833 & 1.03562 & 0.310686 & 0. & 0. & 0. & 0. & 0. \\
0.372823 & 0.370081 & 1.16507 & 1.29985 & 0.582536 & 0. & 0. & 0. & 0. & 0. \\
0.660207 & 0.333578 & 0.349521 & 0.582536 & 0.349521 & 0. & 0. & 0. & 0. & 0. \\
0.372823 & 0.555964 & 0.296514 & 0. & 0. & 0. & 0. & 0. & 0. & 0. \\
0. & 0.294334 & 0.463303 & 0.26163 & 0. & 0. & 0. & 0. & 0. & 0. \\
0. & 0. & 0.228926 & 0.381544 & 0.228926 & 0. & 0. & 0. & 0. & 0. \\
0.198403 & 0. & 0. & 0. & 0. & 0. & 0. & 0. & 0. & 0. \\
0. & 0.125307 & 0. & 0. & 0. & 0. & 0. & 0. & 0. & 0. \\
0. & 0. & 0.0765765 & 0. & 0. & 0. & 0. & 0. & 0. & 0.
\end{pmatrix}
$$

$$(15.25)$$

We can take this power matrix, and simply multiply it to the initial population vector, to obtain the population distribution at the seventh time frame. Verify this result by comparing Equation 15.26 to Equation 15.24.

$$
\begin{pmatrix}
1.16507 & 1.45736 & 1.23501 & 0.775632 & 0.308401 & 0. & 0. & 0. & 0. & 0. \\
0.390641 & 1.16507 & 1.53833 & 1.03562 & 0.310686 & 0. & 0. & 0. & 0. & 0. \\
0.372823 & 0.370081 & 1.16507 & 1.29985 & 0.582536 & 0. & 0. & 0. & 0. & 0. \\
0.660207 & 0.333578 & 0.349521 & 0.582536 & 0.349521 & 0. & 0. & 0. & 0. & 0. \\
0.372823 & 0.555964 & 0.296514 & 0. & 0. & 0. & 0. & 0. & 0. & 0. \\
0. & 0.294334 & 0.463303 & 0.26163 & 0. & 0. & 0. & 0. & 0. & 0. \\
0. & 0. & 0.228926 & 0.381544 & 0.228926 & 0. & 0. & 0. & 0. & 0. \\
0.198403 & 0. & 0. & 0. & 0. & 0. & 0. & 0. & 0. & 0. \\
0. & 0.125307 & 0. & 0. & 0. & 0. & 0. & 0. & 0. & 0. \\
0. & 0. & 0.0765765 & 0. & 0. & 0. & 0. & 0. & 0. & 0.
\end{pmatrix}
$$

$$
\cdot
\begin{pmatrix}
1000 \\
0 \\
0 \\
0 \\
0 \\
0 \\
0 \\
0 \\
0 \\
0
\end{pmatrix}
=
\begin{pmatrix}
1165.07 \\
390.641 \\
372.823 \\
660.207 \\
372.823 \\
0. \\
0. \\
198.403 \\
0. \\
0.
\end{pmatrix}
\qquad (15.26)
$$

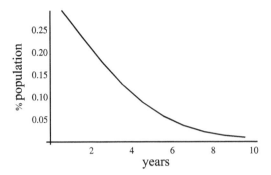

Figure 15.5. Population distribution of our cats at equilibrium.

Population Equilibrium

Again, under the assumption that the population transition matrix stays constant for the indefinite future, we can do another neat trick. We can project forward the population an infinite number of times, that is, we can evaluate $A^\infty.P_0$. The resulting population vector would give us the distribution of population at equilibrium. We can do this by taking the standardized eigenvector associated with the largest real eigenvalue of the Leslie matrix, which I will denote λ_A. Thus, $A^\infty.P_0 = \lambda_A$. The mechanics on how to calculate this eigenvector by hand, along with what exactly eigenvectors and eigenvalues are, lie outside the scope of this course and are better covered in a linear algebra course. For now, we will assume that we have a computer program to calculate this for us.

Using the Leslie matrix representing the population of cats, A, the largest real eigenvalue of A is the first, which is 1.19506. Equation 15.27 shows the eigenvector associated with this eigenvalue, standardized by dividing by the total of all the elements in the eigenvector.

$$
\begin{pmatrix}
-0.659362 \\
-0.524154 \\
-0.394741 \\
-0.280765 \\
-0.187951 \\
-0.117955 \\
-0.0690917 \\
-0.0375795 \\
-0.0188674 \\
-0.00868334
\end{pmatrix}
/ -2.29915 =
\begin{pmatrix}
0.286785 \\
0.227977 \\
0.17169 \\
0.122117 \\
0.0817479 \\
0.0513038 \\
0.030051 \\
0.0163449 \\
0.00820627 \\
0.00377677
\end{pmatrix}
\tag{15.27}
$$

The interpretation is that at equilibrium, we can expect 28.6785% of our cat population to be aged 0, 22.7977% to be 1 year of age, 17.169% to be 2 years of age, and so forth. The structure, given in Figure 15.5, resembles a pyramid shape. This is typical of a population that has high fertility and high mortality. Underdeveloped countries often have this distribution. A population with low fertility and low mortality, which is typical of industrialized countries, will have a more uniform population. The population will not drop off until the very advanced ages.

EXERCISES

For Questions 1 through 6, use the following Leslie matrix, where each state represents a 5-year period.

$$\begin{pmatrix} 0 & .5 & .5 & 0 & 0 & 0 \\ .95 & 0 & 0 & 0 & 0 & 0 \\ 0 & .95 & 0 & 0 & 0 & 0 \\ 0 & 0 & .95 & 0 & 0 & 0 \\ 0 & 0 & 0 & .90 & 0 & 0 \\ 0 & 0 & 0 & 0 & .80 & 0 \end{pmatrix}$$

1. What is the life expectancy at birth, e_0?
2. What is the life expectancy at age 15, e_{15}?
3. With a starting population of 50 people aged 0, what is the total population after two time periods?
4. Of the original cohort of 50, what percentage of the population do they represent after 20 years?
5. Is this population growing, staying stationary, or headed toward extinction?
6. Using matrix multiplication, approximate the population distribution at equilibrium.

For Questions 7 through 10, use the Mathematica demonstration *Population Projection*.

7. Which set of population parameters most heavily influences the shape of the population at equilibrium?
8. If you wanted to greatly prolong the time it took for the population structure to resemble equilibrium, which parameter would you change, and how would you change it? What kind of event could this change represent?
9. The total population increases by 30% in one year. What can you deduce about the shape of the population?
10. Can you design a population, via manipulation of all of the parameters, that will never reach equilibrium?

CHAPTER 16

Evolutionary Game Theory

Consider the following situation (Luce and Raiffa, 1989). Two suspects are taken into custody and separated. The district attorney is certain that they are guilty of a specific crime, but he does not have adequate witness to convict them at a trial. He points out to each prisoner that he has two alternatives: to confess to a crime the police are sure they have done, or not to confess. If they both do not confess, then the district attorney states that he will book them on some very minor trumped-up charge such as petty larceny and illegal possession of a weapon, and they will both receive minor punishment; if they both confess, they will be prosecuted, but he will recommend less than the most severe sentence; but if one confesses and the other does not, then the confessor will receive lenient treatment for turning state's witness, whereas the latter will get "the book" thrown at him. Assume also that neither prisoner can retaliate against an untrustworthy partner.

Table 16.1 shows a set of hypothetical values for years in the pen.

Now consider the following situation. Two roommates share equally in the costs of buying food for their apartment. Each has identical tastes for cognac. To each, a bottle of cognac is worth $30, its price, and no more. Assume that each drinks alone or with his own friends so that each cannot closely monitor the consumption of the other. Simplifying by assuming that each person drinks one or two bottles per month, Table 16.2 shows the outcomes with respect to person A depending on whether they each drink one or two bottles.

The best outcome from A's point of view is to drink two bottles for $45, half the cost of three bottles; the second bottle costs him only $15. The worst outcome is to pay $45 for three bottles (one for him and two for his perfidious roommate) of cognac but to drink only one bottle. Finally, he'd rather pay $30 (half the cost of two bottles) than $60 (half the cost of four bottles). You can see that this situation contains the same dilemma. Since the cost is split, each is tempted to drink more than he usually would because he pays just half the cost of each bottle. But if both do this, both are worse off.

Both of these situations have the same abstract form. Each of two people independently makes a decision to "cooperate" or not to cooperate. Each

Table 16.1.
The prisoner's dilemma

		Prisoner 2's strategy	
		Do not confess (C)	Confess (D)
Prisoner 1's strategy	Do not confess (C)	1 year each	10 years for P1, 3 months for P2
	Confess (D)	3 months for P1, 10 years for P2	8 years each

Table 16.2.
The roommate's dilemma, outcomes for Roommate A

		Roommate B's behavior	
		One drink a day (C)	Two drinks a day (D)
Roommate A's behavior	One drink a day (C)	$30/1 bottle	$45/1 bottle
	Two drinks a day (D)	$45/2 bottles	$60/2 bottles

Table 16.3.
The prisoner's dilemma, abstractly

		Person B's behavior	
		Cooperate	Defect
Person A's behavior	Cooperate	reward (r), reward (r)	sucker payment (s), temptation (t)
	Defect	temptation (t), sucker payment (s)	punishment (p), punishment (p)

faces the following payoffs. The first symbol in each cell is the payoff to Person A, and the second the payoff to Person B.

The key property is that $t > r > p > s$. These conditions ensure that defection, failure to cooperate with one's partner, dominates cooperation as a strategy. It is better to defect than to cooperate no matter what choice the partner makes. But, the result when each person follows his best strategy is an outcome that is not optimal; they could both do better if

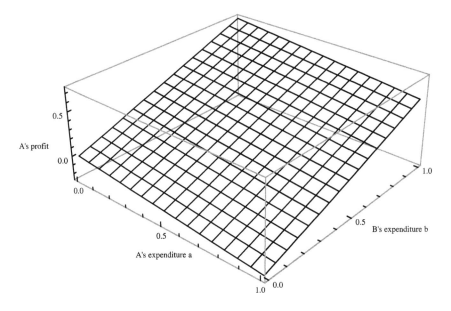

Figure 16.1. Firm A's profit as a function of both firms' expenditures.

they "irrationally" cooperated. When actually allowed to play this game in laboratory experiments, some subjects cooperate and others don't. The interesting question is why any do!

Finally, consider the following example. Suppose that there are two companies, A and B, in an industry that produce an almost identical product (cigarettes or hamburgers). Every dollar spent on advertising produces a dollar and a half worth of new business for the industry as a whole, half of which goes to each firm. Let a and b be the amounts spent on advertising by the two firms. Then the profits P_A and P_B of each firm from advertising are given by the following equations:

$$P_A(a, b) = 1.5\frac{(a + b)}{2} - a$$

$$P_B(a, b) = 1.5\frac{(a + b)}{2} - b$$

(16.1)

Suppose that A and B cannot collude in their advertising strategies because of antitrust legislation. Suppose that A and B are each spending one million dollars on advertising. Each realizes one million dollars in additional profits. Note however that each separately finds that if they reduce expenditures by a dollar they increase their profits by 50 cents. This is true even though if both follow this "rational" strategy neither enjoys the benefits of advertising. Figure 16.1 plots A's profits as a function of its own and B's expenditures on advertising.

Table 16.4.

A game with two pure and one mixed Nash equilibrium pairs

		Person B	
		1	2
Person A	1	3,5	3,3
	2	1,1	4,4

There are clearly innumerable analogous situations involving more than two people. Everyone who values noncommercial radio and television benefits if it thrives, but because those who do not contribute cannot be excluded from its benefits, everyone will have an incentive to be a free rider, enjoying the benefits without contributing. Small firms can enjoy the benefits of legislative lobbying efforts by larger firms without themselves contributing. In both cases a *public good* is underfunded.

However, in this chapter we will confine ourselves to two-person prisoner's dilemma situations. Note that the situation would be quite different if the two prisoners could come to a binding agreement to cooperate with one another. If both had recourse, for example, to a criminal gang that would punish anyone who ratted on his fellow prisoner or if the competing firms could come to a contractual agreement of advertising spending, the outcome would be quite different. Because the actors cannot come to a binding agreement, we will be able to make use of a concept from game theory, the *Nash equilibrium*, named after its discoverer, the great mathematician John Nash. Nash equilibrium is a set (in this case a pair) of strategies whose outcomes cannot be improved upon for each player were she to act independently of the other. The Nash equilibrium is an appropriate solution for *noncooperative games*, in which players cannot come to binding agreements. It's clear that both players defecting in a prisoner's dilemma (PD) situation is a Nash equilibrium pair of strategies. Moreover, it is the only equilibrium pair, as you should easily be able to verify.

John Nash proved that every noncooperative game with a finite number of strategies has at least one Nash solution. Solutions come in two types: *mixed* and *pure*. Pure solutions exist when a player chooses one of his strategies 100% of the time; mixed strategies exist when a player must randomly distribute his choices over a set of his strategies. Let's look at a simple example, given by Table 16.4, in which both pure and mixed strategy solutions exist.

Two of the Nash equilibriums are obvious: the solutions in which both players use Strategy 1 or both players use Strategy 2. The third mixed solution is less obvious. You should convince yourself that if Player A plays Strategy 1 with a probability of 3/5 and B plays 1 with a probability of 1/3, neither player will have a unilateral incentive to change her strategy. Thus, this is a mixed Nash equilibrium.

There are two demonstrations that you can use to familiarize yourself with the concept of a Nash solution. The demonstration labeled *Nash Equilibrium in* 2 × 2 *Mixed Extended Games* gives you the pure and mixed solutions in two-player games in which each player has two strategies.

The PD is not an obscure pedantic topic. It has become a topic of interest in all the social sciences and helped provoke a revolutionary change in our understanding of evolution. The problem is the presence of mutually beneficial cooperation in animals and humans when there are boundless opportunities to exploit the cooperation of others through free riding, letting others do the work while not contributing oneself. Those who defect in the PD earn more. In an evolutionary setting, their greater profits should make it possible for them to leave more progeny, and in humans their greater success should lead to more imitation. Why are human (and animal) societies not a war of all against all? Each of us may have our own opinions on this, but the PD situation allows for careful experimentation and development of models to account for the prevalence of cooperation.

ITERATED PRISONER'S DILEMMA

An *iterated* PD is one that is played a fixed number of times between the same pair of players. Say that two players know in advance that they are going to face an identical PD situation a fixed number of times T. One would think that this would encourage cooperation, especially in the early rounds, when a defection by one player could be punished in future rounds. Paradoxically, this is not the case. Defection in the last round cannot be punished, so both players should defect in the last round. But then, if both players are certain of defecting in the last round, then cooperation in round $T - 1$ cannot be rewarded. Working backward, both players should defect in all rounds, and, unfortunately, this is a Nash equilibrium outcome.

Suppose, however, that there is no fixed number of rounds in the game but instead a probability α after each game that there will be another round and a probability $1 - \alpha$ that there will never be another occasion in which the two prisoners face a PD. The expected number of games is $\frac{1}{1-\alpha}$, but there is no certainty; the last game may be the last one, or there may be many, many more. Since there is no known last game, the logic described in the previous paragraph does not apply.

Robert Axelrod, a political scientist, famously distributed an announcement to game theorists and mathematicians asking for strategies that would compete in a tournament to establish the most successful strategies in iterated two-person PD games. In the computer simulations, each submitted strategy was to play each other strategy 200 times. To eliminate end effects, the strategies could not make use of this fact. In two different tournaments, with 14 and 62 entrants, in which contributors to the second tournament knew the results of the first, a very simple strategy, called TIT-FOR-TAT (TFT), won; it accumulated more points than any strategy. TFT cooperated with each other strategy on its first move and then, on future moves, repeated what its partner had done on the last move. Axelrod attributes the success of TFT to four features.

1. It is nice. It cooperates on the first move, thus avoiding retaliation by others.
2. It's retaliatory. It punishes and does not reward defection by others.
3. It's forgiving. Thus, it is able to tolerate some defection by others without getting into mutually harmful cycles of retaliation.
4. It's simple. Thus, it is easy for partners to recognize that they cannot fool it with complex strategies.

EVOLUTIONARY STABILITY

Even more significantly, TFT was the winner of an evolutionary tournament. This simulation was intended to mimic biological evolution over generations of organisms that face evolving ecologies. Axelrod used all the strategies that had been submitted. At the end of each stage of the evolutionary game, strategies increased or decreased in number according to their success in that stage. Thus, the composition of the population was changing over rounds. Maladaptive strategies and strategies that depended for their success on badly adapted strategies gradually disappeared and the environment became more and more challenging. Again, TFT was the winner.

Axelrod's work and the research that it inspired have profoundly affected the study of both evolutionary biology and game theory. Evolutionary biology has developed a method and set of concepts for making the study of evolution more rigorous. Game theory has been able to give up its assumption that individuals are rational and capable of complex calculation.

The key concept that has been developed is that of an evolutionarily stable strategy (ESS). A strategy is evolutionarily stable if a population consisting of units using that strategy cannot be successfully invaded by any units using a different strategy. A new element following a different strategy is going to face a homogeneous population containing only the strategy under consideration. Thus, whether it thrives will depend on whether it is more profitable against the old strategy than the old strategy is against itself.

Consider the following example. Suppose a population consists exclusively of cooperators. A single defector enters. Her average gain will be t, and the average gain of the cooperators will be r. since $t > r$, the defector will thrive and grow in frequency if growth is proportional to rewards. Therefore, unconditional cooperation is not an ESS.

Consider, on the other hand, a population consisting of only unconditional defectors. No entering strategy can make any more than p. Therefore, no entering strategy can invade a population of defectors.

A variety of simulation models have been adopted to model the evolution of behavior. These models have served two somewhat different purposes. One purpose is to model the actual evolution of behavior patterns among animals. For example, the evolution of cooperation or altruism is a puzzle because it often demands a sacrifice of short-term interests. Second, an evolutionary logic is often used as a substitute for the objectionable assumption of hyper-rationality required by mathematical game theory.

What distinguishes an iterated game is that actors can react to others on the basis of their history of past behavior. Many models have been developed in which actors develop reputations that affect how others, even those they have not yet interacted with, treat them. We will focus in this chapter on one kind of model: actors keep track of how others have treated them and react accordingly. Even more specifically, in the simulation *Evolutionary Prisoner's Dilemma Tournament* each player is limited to remembering how he's been treated in the last game he played with an opponent/partner. Each player also has a choice about how he will treat others with whom he has not interacted. The combination of these choices leads to eight different possible strategies, each described by three letters: what the person does on the first move if he has no history with his partner, what he does if the person has cooperated on the last move, and what he does if the person has defected on the last move. Table 16.5 shows the eight possibilities and gives brief characterizations of them. It also lists a few strategies that cannot be described in this simple way.

The demonstration allows the user to control the initial proportions of the eight strategy types and the number of times pairs of actors are matched. At the maximum number of iterations, 3,000, each of the 100 players will interact with each other player an average of $3{,}000/100 = 30$ times. The demonstration also allows for some control of the values s, r, p, and t. The advantage to strategies that adjust to the behavior of others, like TFT, will depend on the number of iterations. The success of a strategy determines its representation in the next generation.

To make these ideas more concrete before we examine the demonstration, let's look at a tournament involving just two strategies, TFT and unconditional defection (D). Let α be the probability that a game between two players is repeated. $\frac{1}{1-\alpha}$ is the expected number of games between any two players. Table 16.6 shows the expected gains to the row player as she plays an opponent with using TFT or D.

Let q be the proportion of TFT players in the population. q and $1-q$ are the probabilities that any player's partner will be TFT or D. The expected

Table 16.5.
Strategies in the iterated PD game

Strategy	Description
CCC	Unconditional cooperation
CCD	Tit-for-tat
CDC	Cooperates on first move, then does opposite of what partner did
CDD	An unconditionally defecting strategy that cooperates on the first move
DCC	An unconditionally cooperative strategy that defects on the first move
DCD	A tit-for-tat strategy that defects on the first move
DDC	Defects on the first move, then does opposite of what partner did
DDD	Unconditional defection
Trigger	Cooperate until the other defects, then never cooperate
Two-tits-for-a-tat	Defect only when the other has defected for two consecutive moves
Random	Cooperate but defect ocassionally

Table 16.6.
Expected earnings for a row player in TFT versus unconditional defection

	TFT	D
TFT	$r/(1-\alpha)$	$s + \alpha p/(1-\alpha)$
D	$T + \alpha p/(1-\alpha)$	$p/(1-\alpha)$

rewards for TFT and D players are thus,

$$R_{TFT} = \frac{qr}{1-\alpha} - (1-q)\left(s + \frac{\alpha p}{1-\alpha}\right)$$

$$R_D = q\left(t + \frac{\alpha p}{1-\alpha}\right) + \frac{(1-q)p}{1-\alpha}$$

(16.2)

For the sake of simplicity, let us assume that the values of t, r, p, and s take their default values in the simulation: 6, 4, 2, and 0. Remember that $\frac{1}{1-\alpha}$ is the average number of interactions between a pair. Also, the greater q, the greater the opportunities for TFT strategists to interact beneficially with each other. So, it should be clear that for high values of α and for q, the TFT strategy will outperform the D strategy in a population in which the two are randomly mixed. Figure 16.2 shows the average advantage of the TFT strategy when that value is positive. It shows that high values of α are required before the TFT strategy can thrive against the D strategy.

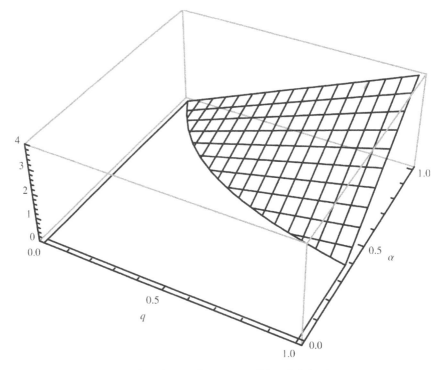

Figure 16.2. When TFT does better than unconditional defection.

For the region outlined in Figure 16.2 TFT will grow in a population at the expense of D. These values will, of course, be modified as the values of the parameters t, r, p, and s change.

Chapter Demonstrations

- *Evolutionary Prisoner's Dilemma Tournament* enables you to conduct your own evolutionary tournaments using a small set of strategies
- *Nash Equilibrium in 2 × 2 Mixed Extended Games* shows the pure and mixed Nash equilibriums for a set of two-person games where each player has two choices.

EXERCISES

1. Select the Nash pure equilibrium pair in the following two-person games, if there is one. The first number in each cell is the payoff to the row player; the second number is the payoff to the column player. Circle the pure Nash equilibrium, if any.

1, 1	20, −1
−1, 20	10, 10

1, 1	0, 0
0, 0	10, 10

6, 30	4, 42	39, 6
24, 17	4, 9	21, 34
14, 24	19, 25	24, 49

2. If there are 2,000 iterations in a PD game, to what value of α does this correspond?

3. Verify that when α is near .06 and the number of iterations is 2,000, both TFT and D do approximately equally well. What should be the effect of changing each of the four parameters t, r, p, and s on the relative success of TFT and D?

4. In the demonstration, all strategies start out with an equal frequency. If TFT starts out at a sufficiently low value, does it still win in the long run? If DDD starts out at a sufficiently high value, does TFT still win?

5. Try a simulation with all eight strategies if which there are 10 iterations per round. Which strategies win? Now try 2,000 iterations. What is the winning strategy after many generations?

CHAPTER 17

Power and Cooperative Games

Consider the following romantic triangle. Andy and Charlie are both in love with Barbara, and she likes both of them. Barbara likes to go to plays, which Andy and Charlie don't particularly care for (Andy likes professional wrestling, and Charlie likes to bicycle). Andy and Charlie both end up going to a lot of plays they don't particularly like and even saying they enjoy them because they want to win Barbara. Barbara clearly has power in this situation.

We could diagram this situation as a graph with three vertices and two edges, given in Figure 17.1.

Now suppose that Andy has a job during the summer and meets two young women, Debbie and Elisa, whom he likes and who like him. Now the situation can be represented by Figure 17.2.

Now the power relations are quite different. Because Andy has two other alternatives to Barbara, Barbara no longer has much power over Andy. And because Andy is no longer competing with Charlie, Barbara also has less power over Charlie than she used to. Andy is clearly the big winner from the change. He now has three women interested in him, two of whom are interested in him exclusively.

The dynamics of power that we described are not limited to dating situations. In Figure 17.1 Andy and Charlie could be two potential buyers for Barbara's car. Since they are competing with one another, the price should drop. But what if, after looking at Craigslist, Andy discovers two other cars for sale that he likes equally well and for which there are no other buyers. Or, alternatively, suppose that Andy and Charlie both want Barbara as their regular racquetball partner. She can determine when and where she wants to play. But, suppose that Andy finds two others players, Debbie and Elisa, who want him as regular partners. Then, Barbara loses power in determining the conditions under which she plays, and Charlie and Andy gain power.

In this chapter we will be looking at models of the following type of idealized and tractable situation. We will look at it because there is a great deal of experimental and theoretical work on this type of situation and because it allows us to introduce some concepts and tools from *game theory*, a useful and elegant theory of rational choice developed by

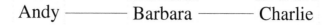

Figure 17.1. A romantic exchange network with three nodes.

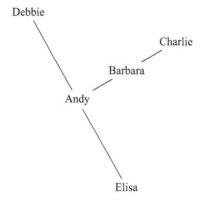

Figure 17.2. A romantic exchange network with five nodes.

mathematicians and economists.

1. There is a certain value to transactions between pairs {i,j} of people. The value of the transaction can be divided in any way between the members of each pair.
2. Individuals must choose one person, or a limited number of people, with whom to transact. Some individuals may have to select among potential partners while others risk having no partner at all.

Notice that some situations fit this model imperfectly and others not at all.

1. This model fits the dating situation very imperfectly. Although there are norms and jealousies that may make it impossible, nothing prevents Barbara from dating Andy and Charlie simultaneously. Even more seriously, there may be asymmetries; Barbara may enjoy Andy's presence more than Andy enjoys Barbara's. There will be values in the relationship that cannot be transferred or divided between the partners; Barbara's appreciation of Andy's intelligence cannot be given to Andy.
2. On the other hand, the situation in which individuals are buying and selling automobiles may fit almost perfectly. There is a profit to the transaction, the dollar difference between what the buyer is willing to pay and what the seller is willing to accept, that can be divided in any way between the two of them depending on the price for which the car is sold.

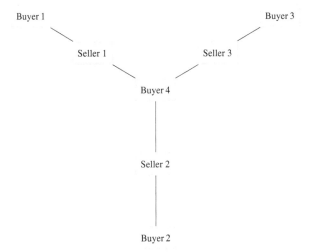

Figure 17.3. An exchange network with seven nodes.

It is unfortunate that the kinds of situations that fit this model best are situations that interest economists rather than sociologists, but, as we shall see, putting the sociological situations into this Procrustean bed does provide insights. Moreover, it may not be at all obvious who has the power in a larger and more complicated network, and having a mathematical model, even if it does not fit perfectly, can help us understand complex situations where our intuition and insight fail. Consider, for example, the following situation in Figure 17.3.

Suppose that automobiles are being bought and sold and the lines connect pairs that can engage in a profitable exchange (the buyer would be willing to pay more than the seller's minimum price). Suppose that each seller has but one vehicle to sell and each buyer desires only one vehicle. Suppose also, because we are focusing on the effect of the network structure, that all the exchanges have the same possible total value, $3,000.

It might seem as if Buyer 4 is in an ideal situation. He has three automobiles to choose from, and if Buyers 1, 2, and 3 did not exist he would indeed be in a very good situation. The three sellers would all be competing for his business. He would enjoy the advantages of a monopsony, in the language of economists—many sellers and just one buyer. However, there are other buyers apart from him. He is in the unfortunate situation of competing with buyers who themselves have no alternatives for the cars they desire. So, it's the three sellers who are in the best position.

In Figure 17.4 the inner pair, Seller 1 and Buyer 2, are also in a favorable position, but not quite as favorable. Each is faced with one person with no alternatives but not two or more who bid against one another.

Buyer 2 ——— Seller 1 ——— Buyer 1 ——— Seller 2

Figure 17.4. An exchange network with four nodes.

You could substitute other more sociological but less perfectly realized situations. Imagine an unmarried woman dating three married men. She does not realize the full advantages of having three suitors. Or, imagine an expert in many areas who does not realize the full power and status advantages of his expertise because there are alternative sources of expertise in each of his specialties.

In this chapter we will examine two solutions to these situations that have been developed by mathematicians and tested by sociologists: the *kernel* and the *core*. These are solutions developed within an area of mathematics called somewhat misleadingly game theory, because its usefulness goes well beyond the analysis of games. Game theory is concerned with all situations that involve strategic interaction between *rational* actors. Rational actors are assumed to act in their own best interests. Actors have interdependent interests (what is best for each is likely to depend on what others are doing), and what happens when all are simultaneously reacting to what others are doing can be quite complex.

The two solutions we are going to examine both apply to *cooperative* games in *characteristic value* form. In cooperative games actors can arrive at binding agreements, as distinct from *noncooperative games*, when they cannot. The difference is profound. Imagine what business activity would be like if contracts were impossible to enforce. War is the ultimate noncooperative game, during which even international treaties about weapons or treatment of prisoners may be broken.

In a special kind of cooperative game, one that can be cast in characteristic value form, every subset of actors has a value, and this value can be allocated between the actors in any way. To take a simple example, suppose three people with different amounts to invest can pool their money and make an investment that will have a higher payoff the more they invest. What they must agree on is the division of the payoff.

More specifically, let S be a set of players. Remember, from Chapter 2, that 2^S, the power set, is the set of all subsets of S. The characteristic value function v is a function from the power set to the real numbers; if T is a subset of S, then $v(T)$ is the amount that the subset can achieve and that it can divide between its members in any way it chooses. For example, suppose we have three investors A, B, and C with 10, 10, and 20 units to invest and that investments of at least 10, 20, and 30 units earn 5%, 7%, and 9%, respectively. The characteristic value function would be as follows (where the empty set ø is added for the sake of completeness).

$$v(\emptyset) = 0 = 0$$

$$v(\{A\}) = 10 + 10 \times .05 = 10.5$$

$$v(\{B\}) = 10 + 10 \times .05 = 10.5$$

$$v(\{C\}) = 20 + 20 \times .07 = 21.4$$

$$v(\{A, B\}) = 10 + 10 + .07 \times (10 + 10) = 21.4$$

$$v(\{A, C\}) = 10 + 20 + .09 \times (10 + 20) = 32.7$$

$$v(\{B, C\}) = 10 + 20 + .09 \times (10 + 20) = 32.7$$

$$v(\{A, B, C\}) = 10 + 10 + 20 + .09 \times (10 + 10 + 20) = 43.6$$

The vector $x = \{x_1, x_2, x_3, \ldots, x_N\}$ represents that final payoffs to the N participants. Suppose that all three invested all their money and split the rewards equally. Then, $43.6/3 = 14.53$ and $x = (14.53, 14.53, 14.53)$. Note how this solution seems unfair, especially to C, who can earn 21.4 without any partners at all.

In this chapter we will be focusing on network exchange games, much studied in social psychology, in which the only valuable transactions involve two individuals and the exchange possibilities can be represented as a network. Let's look at the exchange network in Figure 17.1, consisting of Andy, Barbara, and Charlie, as if it were a cooperative game in characteristic value form. Each person wants to spend exclusive time in an activity he or she enjoys with one member of the opposite sex. We assume that any couple can divide their time between activities that each of them especially enjoys and that two is a company but three's a crowd, the second boy adding nothing to the enjoyment possible for either boy-girl pair.

$$v(\emptyset) = 0$$

$$v(\{A\}) = 0$$

$$v(\{B\}) = 0$$

$$v(\{C\}) = 0$$

$$v(\{A, B\}) = 2$$

$$v(\{A, C\}) = 0$$

$$v(\{B, C\}) = 2$$

$$v(\{A, B, C\}) = 2$$

THE KERNEL

The kernel can be seen as a way of representing mathematically a conception of power that has existed in social psychology for a long time: power-dependence theory (Emerson, 1976). One person is dependent on another to the degree that the former receives valuable rewards from the latter that are not available elsewhere. A child is heavily dependent on her parents. An employee is heavily dependent on his employer if jobs are scarce. Of course, the employer is also dependent to some extent on an employee with valuable skills and experience, but differences in dependence lead to differences in power. The more dependent member will tend to be more accommodating to the wishes of the more dependent. Dependence is affected not only by the value of the reward but also by the availability of the reward from alternative sources. No matter how valuable one person is to another, there is no dependence if some equally or more valuable alternative source of the same reward is available.

Consider, for example, the well-known but fictional Godfather (Puzo, 1970). He does valuable favors for others, but none of the favors can be performed by other legitimate sources. He lends money to one person who cannot get a loan from a bank. He punishes someone's daughter's wealthy rapist after the rapist has gotten off scot-free because his family could afford an expensive lawyer. The debtors would all prefer the legitimate sources because they know that the Godfather will later ask them to return the favor by performing some kind of illegal activity themselves.

The second principle of power-dependence theory is that power differences tend to disappear over time. Once the indebted person performs a service for the person to whom she is indebted she increases the dependency of the latter by performing a valuable service. A popular girl may find that only one boy, not necessarily the best looking or personable, is willing to indulge her taste in going to NASCAR events.

Now for a technical definition of the *kernel*, the game theory concept that has proven useful in the study of exchange networks. Let S be a subset of actors in a cooperative game in characteristic value form and let x be a set of payoffs. The *excess* $e(S, x)$ of a coalition S is the difference between its value $v(S)$ and the total payoffs received by the members of the coalition.

$$e(S, x) = v(S) - \sum_{i \in S} x_i \qquad (17.1)$$

A set of players S can, acting together, earn more or less than the values they are receiving from other coalitions. Suppose, for example, that all three investors form a coalition and decide the split the profits of 43.6 equally so that each earns 14.53. Then the other nonempty coalitions will have positive or negative excesses.

$$e(\emptyset, x) = v(\emptyset) - 0 = 0 - 0 = 0$$

$$e(\{A\}, x) = v(\{A\}) - x_A = 10.5 - 14.53 = -4.03$$

$$e(\{B\}, x) = v(\{B\}) - x_B = 10.5 - 14.53 = -4.03$$

$$e(\{C\}, x) = v(\{C\}) - x_C = 21.4 - 14.53 = 9.87$$

$$e(\{A, B\}, x) = v(\{A, B\}) - x_A - x_B = 21.4 - 14.53 - 14.53 = -7.66$$

$$e(\{A, C\}) = v(\{A, C\}) - x_A - x_B = 32.7 - 14.53 - 14.53 = 3.64$$

$$e(\{B, C\}) = v(\{B, C\}) - x_B - x_C = 32.7 - 14.53 - 14.53 = 3.64$$

$$e(\{A, B, C\}) = v(\{A, B, C\}) - x_A - x_B - x_C = 43.6 - 14.53 - 14.53 - 14.53 = 0$$

Coalitions with positive surpluses do not earn as much with an equal division as they could guarantee themselves by acting together. The set {A,C} are earning only 29.06, whereas if they acted together without B they could earn 32.7, V({A,C}), between them. C could earn 9.87, more, V({C}) – 14.53, by himself than in this equal division. On the other hand, {B,C} as a pair do better in this coalition of all the players (with an equal division) than they could on their own.

A division of the profits within a coalition is in the kernel if all pairs of members are equally dependent on one another; that is to say, in no pair is one member of the coalition more dependent on the other. In every member of a pair each assesses his position in relation to the other. Each asks himself, how much worse off or better am I than in the best alternative coalition available to me? If all members of a coalition are equally better off within it than with respect to their best alternative, they should all be equally loyal members. More formally, for every pair of players i and j,

$$\max_{\substack{i \in S \\ j \notin S}} e(S, x) = \max_{\substack{j \in T \\ i \notin T}} e(T, x) \qquad (17.2)$$

The rationale for the kernel is psychological, not rational. It is based on the power-dependence principle. The more dependent actor in the relationship has less power. The more dependent actor will be more accommodating and the less dependent more assertive. Adjustments will continue until both are equally dependent on each other in relation to what they can receive from alternative relationships.

To solve the problem of the three investors more systematically, let's look at Table 17.1. The second column gives the value of the coalition. The third column gives what those coalition members earn under an equal division. The fourth column shows the excess, the amount those coalition members would earn if they acted alone instead of splitting the rewards equally.

Table 17.1.
Coalition excesses under an equal division by three investors

Coalition S	$V(S)$	$\sum_{i \in S} x_i$	$V(S) - \sum_{i \in S} x_i$
$\emptyset$	0	0	0
$\{A\}$	10.5	14.53	−4.03
$\{B\}$	10.5	14.53	−4.03
$\{C\}$	21.4	14.43	6.87
$\{A, B\}$	21.4	29.06	−7.66
$\{A, C\}$	32.7	29.06	3.64
$\{B, C\}$	32.7	29.06	3.64
$\{A, B, C\}$	43.6	43.59	0

Table 17.2.
Coalition excesses in the kernel

Coalition S	$V(S)$	$\sum_{i \in S} x_i$	$V(S) - \sum_{i \in S} x_i$
$\emptyset$	0	0	0
$\{A\}$	10.5	10.76	−.26
$\{B\}$	10.5	10.76	−.26
$\{C\}$	21.4	22.07	−.67
$\{A, B\}$	21.4	21.52	−.13
$\{A, C\}$	32.7	32.83	−.13
$\{B, C\}$	32.7	32.83	−.13
$\{A, B, C\}$	43.6	43.59	0

This outcome is clearly not the kernel. Consider C's dependence on A. C belongs to two coalitions that exclude A, {C} and {C,B}, with excesses 6.87 and 3.64, respectively. The maximum of these excesses is 6.87. C could earn up to 6.87 more points by acting on her own, excluding A. On the other hand, A belongs to two possible coalitions without C, {A} and {A,B}, with excesses of −4.03 and −7.66, respectively. Taking the maximum of these, A would lose 4.03 if he left the {A,B,C} coalition with an equal split. Thus A is much more dependent on C than C is on A; their maximum excesses are not equal. This division is not the kernel.

On the other hand, consider the distribution of rewards for the coalition {A,B,C} in which A and B earn 10.76 and C earns 22.07. See Table 17.2. The two coalitions A belongs to that C does not are {A} and {A,B} with excesses of −.26 and −.13. The best that A can do independent of C is to suffer a loss of .13. C belongs to two coalitions without A, {C} and {B,C}, with excesses of −.67 and −.13. The best C can do independent of A is also to suffer a possible loss of .13. They are thus equally dependent upon one another, as are B and C and A and C. Therefore, this allocation is the kernel.

Let's see how this logic plays out in the network in Figure 17.4, where it seemed plausible that Seller 1 and Buyer 1 enjoyed some advantage. Suppose that each exchange is worth a potential $3,000 (the difference

Table 17.3.

The kernel for Figure 17.4 if all relations are worth $3,000

Coalition S	$V(S)$	$\sum_{i \in S} x_i$	$V(S) - \sum_{i \in S} x_i$
$\emptyset$	0	0	0
$\{B_2\}$	0	1000	−1000
$\{S_1\}$	0	2000	−2000
$\{B_1\}$	0	2000	−2000
$\{S_2\}$	0	1000	−1000
$\{B_2, S_1\}$	3000	3000	0
$\{B_2, B_1\}$	0	3000	−3000
$\{B_2, S_2\}$	0	4000	−2000
$\{S_1, B_1\}$	3000	4000	−1000
$\{S_1, S_2\}$	0	3000	−3000
$\{B_1, S_2\}$	3000	3000	0
$\{B_2, S_1, B_1\}$	3000	5000	−2000
$\{B_2, S_1, S_2\}$	3000	4000	−1000
$\{B_2, B_1, S_2\}$	3000	4000	−1000
$\{S_1, B_1, S_2\}$	3000	5000	−2000
$\{B_2, S_1, B_1, S_2\}$	6000	6000	0

between what the buyer is willing to pay and what the seller is willing to accept). Let x = {xB$_2$, xS$_1$, xB$_1$, xS$_2$} and suppose that B$_2$ and S$_1$ exchange as do B$_1$ and S$_2$. Table 17.3 shows the outcomes if the middle pair S$_1$ and B$_1$ both earn 2/3 of the profit in their transactions: $2,000.

Consider the power-dependence relation between B$_2$ and S$_1$. The greatest excess among all the coalitions B$_2$ belongs to that S$_1$ does not is −$1,000, for the coalition {B$_2$}. The best he can do independent of S$_1$ is to lose $1,000. The coalition with the greatest excess among all those S$_1$ belongs to that B$_2$ does not is {S$_1$, B$_1$}, which B$_1$ would want at least $2,000 to join, leaving $1,000 less for S$_1$ than he is now earning. Therefore, they are equally dependent on each other: both would lose $1,000 by joining their best alternative coalition, the one with the greatest excess. This solution is, therefore, the kernel.

Now consider a slightly more complicated situation, one in which the trades are not all of equal value. Suppose that the B$_2$-S$_1$, S$_1$-B$_1$, and B$_1$-S$_2$ trades have values of $3,000, $1,000, and $2,000, respectively. We will show that the relationship between the middle positions is no longer valuable enough for them to enjoy any advantage. Suppose that B$_2$ exchanges with S$_1$ and B$_1$ with S$_2$, as before. We will show that the kernel is an equal division between members of the exchanging pairs. Let's look at Table 17.4.

Consider B$_2$ and S$_1$. B$_2$ belongs to just one coalition not involving S$_1$, {B$_2$}, and he would lose $1,500 by not trading with S$_1$ (the excess for the coalition {B$_2$} is −$1,500). S$_1$ belongs to four coalitions not including B$_2$ (2, 9, 10, and 15), but the maximum excess is also −$1,500; he would have to pay B$_1$ at least $1,000 to leave his current exchange partner, leaving nothing for S$_1$. Therefore, both would lose equally by rejecting each other.

Table 17.4.
The kernel for Figure 17.4 if all relations are worth $3,000, $1,000, and $2,000

Coalition S	$V(S)$	$\sum_{i \in S} x_i$	$V(S) - \sum_{i \in S} x_i$
$\emptyset$	0	0	0
$\{B_2\}$	0	1500	−1500
$\{S_1\}$	0	1500	−1500
$\{B_1\}$	0	1500	−1000
$\{S_2\}$	0	1000	−1000
$\{B_2, S_1\}$	3000	3000	0
$\{B_2, B_1\}$	0	2500	−2500
$\{B_2, S_2\}$	0	2500	−2500
$\{S_1, B_1\}$	1000	2500	−1500
$\{S_1, S_2\}$	0	2500	−2500
$\{B_1, S_2\}$	2000	2000	0
$\{B_2, S_1, B_1\}$	3000	4000	−1000
$\{B_2, S_1, S_2\}$	3000	4000	−1000
$\{B_2, B_1, S_2\}$	2000	3500	−1500
$\{S_1, B_1, S_2\}$	2000	3500	−1500
$\{B_2, S_1, B_1, S_2\}$	5000	5000	0

The same holds for B_1 and S_2; they are equally dependent on one another. Therefore, an equal division within each of the two dyads is the kernel.

THE CORE

The second game-theoretic mathematical model we will examine for cooperative games in characteristic value form is based on a different assumption. It is simply that no settlement will occur in which the value of the coalition is greater than what its members are receiving. Technically, no coalition should have a positive excess (as defined by Equation 17.1). Why should any set of individuals accept less than what they can earn if they act in concert? In the situation described in Figure 17.4, where all coalitions are of size 1 or 2, this means that the following conditions should hold:

$$x_{B_2} \geq 0$$
$$x_{S_1} \geq 0$$
$$x_{B_1} \geq 0$$
$$x_{B_2} \geq 0 \tag{17.3}$$
$$x_{B_2} + x_{S_2} \geq 3000$$
$$x_{B_1} + x_{S_1} \geq 3000$$
$$x_{B_1} + x_{S_2} \geq 3000$$

The first four conditions state the trivial fact that no participant should lose money, since she can always simply refuse to buy or sell. The only way all these conditions can be met is for Seller 1 to sell to Buyer 2, for Buyer 1 to sell to Seller 2, and for the total profits of Seller 1 and Buyer 1 to be greater than or equal to $3,000; Seller 1 and Buyer 1 together must make at least half the total profit—they might make considerably more.

The demonstration *Exchange Networks* simulates actor behaviors that will invariably produce an outcome in the core. There is some evidence that it also approximates subjects' behavior in experimental groups. The rule the actors follow is that if they are excluded from an exchange in one round they increase their offer to others in the next by two points and that if they are included in an exchange in one round they lower their offers to others by one point in the next. The simulation allows you to examine the growth of network power in a variety of networks and also to control the behavior of one actor, to see if you can devise a strategy that does better than the default strategy.

The interesting feature both models share is that the outcomes depend in complicated ways on features of the entire network. A small change in one part of the network can have profound implications for distant parts of the network. This is a desirable feature because networks do have this property; a power failure in Seattle can lead to blackouts in Southern California.

Look at the network in Figure 17.3. Both the core and kernel have the same outcome: all the profits should devolve on the three sellers, and the sellers will pay top dollar. However, suppose Buyer 3 disappears. Sellers 1 and 2 are in a much less favorable position, and Buyer 4 now occupies one of the more profitable positions while Seller 3 is demoted to one of the less favorable outcomes.

Chapter Demonstrations

- *Exchange Networks* simulates the behavior of actors in exchange networks who use simple and plausible adjustment rules: if they are excluded in an exchange in one round they raise their offers to others by about two points in the next round; if they are included in an exchange in one round they lower their offer to others in the next round by about a point. This simulated behavior has been shown to reproduce results in the laboratory using actual subjects.

EXERCISES

1. Suppose Andy and Charlie wanted to buy Barbara's automobile (Figure 17.1). Barbara is willing to sell it for $5,000. Charlie is willing to pay

$5,500 while Andy is willing to pay $6,000. What outcomes are in the core? If Andy buys Barbara's car, what price does the kernel suggest?

2. Now suppose that Andy locates two other sellers of the car he wants who are willing to sell it for $4,000. What outcomes are now in the core and in the kernel?

3. Find the core solution for the three investors described in this chapter.

4. Imagine a situation in which three individuals each want an exclusive reciprocated relationship to one but only one of the other two. Suppose, for the sake of simplicity, that this is a cooperative game in characteristic form in which each dyad is worth one unit. What outcomes are in the core?

5. In the demonstration *Exchange Network* there are five illustrative networks. Each actor can come to at most one agreement with another actor to whom she is connected to divide 24 points. In each network you can let each actor simply increase his offers to others in the next round if he is excluded, or you can control the bidding behavior of one of the actors. In the first of the five networks run the simulation many times and observe the relative rewards of Actor 4. Now take control of Actor 4's choices and try to devise a strategy that enables her to make significantly more than she did previously.

6. Now do the same thing with Actor 3 in the first network. Run the simulation to see how much Actor 3 makes, then take control of Actor 3 and try to devise a strategy that is better for him.

CHAPTER 18

Complexity and Chaos

This chapter concerns two new and not fully developed ideas in the mathematical modeling of social science phenomena: chaos and complexity. Both concepts refer not to particular mathematical models but to characteristics of mathematical models. Both also imply limitations on our ultimate abilities to predict the future of human societies no matter how valid our models.

CHAOS

Why are sociologists and other social scientists so unable to predict the future course of society? No one predicted the disappearance of the Soviet Union. No one predicted the current "great recession"; there is no consensus about when it will end. One possible reason is that our theories and our understanding are not very good, but another reason has been offered: chaos. Chaos is a technical mathematical term describing some systems for which accurate predictions of the future are all but impossible. Everyone has heard of the *butterfly effect*, the hypothesis that the movement of the wings of a butterfly in, say, Siberia can have a multiplicative and triggering effect on the weather in California. It has been established that there are inherent reasons for our limited ability to predict the weather far in advance: it is not because the models are bad but because the models are extremely (infinitely) sensitive to small variations in initial conditions, and we can never have perfectly accurate measures of current weather conditions.

There is a well-known story involving the meteorologist Edward Lorenz, one of the researchers in the area of chaos. Lorenz wanted to examine a weather forecast, and to save time he reran part of a forecast by reentering data that had been generated in the middle. To his surprise he found that the forecast was completely different from what it had been. The reason was that he had entered three-digit numbers printed by the computer when the computer itself operated on six-digit numbers. The very small initial differences were enough to produce large differences in the ultimate forecast.

Systems are either deterministic or stochastic. Systems of the former type obey rigid laws that, in theory, enable us to predict the future perfectly. For example, if a population grows at 30% a year, we can predict its future size at year t: $p_t = 1.30^t p_0$, where p_0 is the initial population size. On the other hand, in stochastic systems the unpredictability of the future is built into the model so that we know we can make only probabilistic predictions. Markov chains are examples of stochastic models. Suppose, for example, that in a 10-person group each person had a .30 chance of recruiting a new member in the next year. The expected number of new members is 3, but we can use the binomial distribution to calculate probable results. For example, the probability of getting exactly three new members is .267, and the probability of two new members is .233, and so on. Deterministic models make deterministic predictions and stochastic models make probabilistic predictions.

This distinction (between deterministic and stochastic) is independent of the validity of the model, which refers to the accuracy of its predictions. A deterministic model can be completely invalid. For example, the kernel is a useful model for bargaining situations that may sensitize us to power differences, but it may be highly inaccurate compared to other models in any particular situation. For example, bargaining among members of a family may be determined not by the power-dependence principle but by status. On the other hand, a stochastic model, even though its predictions are never very specific, can be completely valid. Demographic projections cannot be used to predict the number of children any particular woman will have even if the societal predictions are quite accurate.

We are now examining the paradoxical possibility that some deterministic models (regardless of their validity) may be of very limited usefulness in predicting the future. To get a sense for this phenomenon we want to examine three different processes with different of sensitivities to initial conditions. We will be focusing on the predictability of the future in the model and the degree to which the model's outcomes are stable: are there equilibriums toward which the system moves regardless of its initial conditions. A model is more useful for prediction if its outcomes do not depend on accurate measurements of the present.

Stochastic Processes

Stochastic processes are unpredictable because they are not deterministic. Let us see how one representative stochastic process operates. Consider a *random walk* in one dimension. At every step the variable x has a .50 probability of increasing or decreasing by one unit.

$$x_{t+1} = x_t + z_t \qquad (18.1)$$

z is a random variable that takes the values $+1$ or -1 with equal probabilities. Figure 18.1 shows five realizations of a random walk with

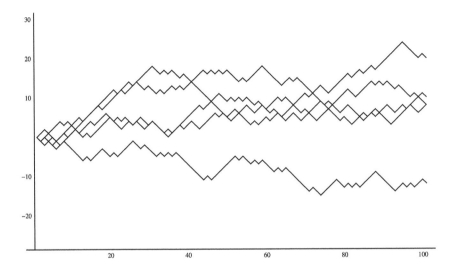

Figure 18.1. Five realizations of a random walk.

100 time periods. The expected value of the sum x_t is zero. Note how the values slowly deviate from that value. The expected deviation grows only as the square root of t, the number of trials. Even though any sequence is unpredictable, the probabilities of ranges of values for x are completely known. This could be a model for any number of things in the real world, including, for example, gambling gains and losses or, in the short run, the stock market.

For another example of the stable nature of stochastic processes, consider the Markov chains we studied in Chapter 14. There are many Markov chains for which the equilibrium long-run distribution of probabilities of states is independent of the initial probability distribution of states. For example, suppose that the transition matrix of identification with two political parties between elections (Figure 14.2) were as follows:

$$\begin{pmatrix} .7 & .3 \\ .1 & .9 \end{pmatrix} \tag{18.2}$$

If the Markov assumptions were true, then, regardless of the initial distribution of a set of voters across the two political parties, in the long run 75% of the voters would be in the second party.

Deterministic Nonchaos

Consider the epidemic model of the sort discussed in Chapter 1.

$$y_{t+1} = y_t + \beta x_t \tag{18.3}$$
$$x_{t+1} = x_t - \beta x_t + \alpha x_t (1 - y_t - x_t) \tag{18.4}$$

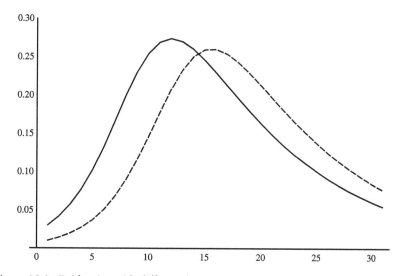

Figure 18.2. Epidemics with different beginning percentages but same parameters.

This is a deterministic model of epidemics. The value x_t is the proportion of the population that is sick and y_t is the proportion that has been sick but has recovered and is immune. The parameter β indexes the contagiousness of the disease, and β measures the speed with which those who are sick recover. Figure 18.2 shows the predicted courses of two epidemics, each with $\alpha = .20$ and $\beta = .025$ but with different starting points: 5% and 1% of the population being sick initially. Note that the two curves are not radically different: the curve starting at 1% is almost a displacement of the curve starting at 1%. The final outcome is not sensitively dependent on the initial starting point.

Many processes and models are like this. Small differences in initial conditions lead to the very same equilibrium outcome, slightly different outcomes (our example), or outcomes that very gradually become different from one another over time.

Deterministic Chaos

There are systems in nature where tiny and insignificant initial differences lead quickly (exponentially) to profound and unpredictable differences in outcomes. There is no limit to how small the initial differences need be. Since measurement is always of limited accuracy, this means that long-range predictions in such systems are impossible, no matter how much is known about the system. Models of the weather, gravitational models involving three or more orbiting bodies, and a number of other systems are known to be chaotic.

The simplest chaotic system involves the logistic curve generating a sequence of values according to a deterministic rule.

$$x_{t+1} = \mu x_t (1 - x_t) \tag{18.5}$$

The demonstrations *Classic Logistic Map* and *Logistic Sequence Sensitivity* can be used together to examine the properties of this sequence.

The parameter μ lies between 0 and 4. For values below 3 the logistic map behaves in an unexceptional manner; the sequence converges to a single value, regardless of the first starting value x_0. At $\mu = 3$ the outcome changes; the results alternate between two values regardless of the starting value. At approximately 3.449 the cycle doubles to four alternating outcomes. The number of outcomes then multiplies more and more rapidly until there is chaos: there are no cycles at all and the results are completely dependent on the starting value. The demonstration *Classic Logistic Map* shows these chaotic regions for μ.

Select a chaotic value for μ (some value near 4.00) and use the demonstration *Logistic Sequence Sensitivity* to explore the sensitivity to starting value. With this demonstration you can examine the path of a sequence and another sequence with a slightly different initial value for using the slider labeled "Δ," creating a second sequence whose starting value differs slightly from that of the first sequence. You will see that the difference in the two sequences is initially quite small until it explodes exponentially—as a baseline, two randomly chosen numbers in the interval between zero and one will differ, on the average, by 1/3.

Although the concept is intriguing, at present there are no useful models in sociology that are chaotic. Most known chaotic models consist of three or more nonlinear differential equations, and sociology does not yet have models of this level of sophistication. Someday it will, and the student of mathematical sociology should be aware of this concept and be able to capitalize on it when sociological models become more mature.

COMPLEXITY

Complexity is another fashionable term in the social sciences today. It is used in two ways. The first, and less controversial, use is synonymous to emergence. It is the recognition that sets of interacting actors or units can have unexpected macro-level properties. We've already seen this in a number of different contexts: nodes in a network are suddenly connected in one large component if the average number of connections per node passes 1.00 (Chapter 1); rational actors produce outcomes that are suboptimal for all in prisoner's-dilemma-like situations (Chapter 16); individuals who simply are influenced by their friends' preferences produce clique structures (Chapter 13); individuals using preferential attachment to form connections when they join new networks produce scale-free networks (Chapter 12).

Moreover, sometimes these emergent properties of systems of independent yet interconnected units can be so complicated that there are no exact mathematical solutions and computer simulations may be the only way to get a handle of emerges. Another textbook could be written on using computer simulations to study complex social system properties.

However, complexity has a second, more controversial meaning. It hints at the possibility that there are underlying rules and processes that occur in all complex systems, regardless of their content. In this chapter we will focus on one possible regularity that has gotten lots of attention, the existence of power-law distributions in many different complex systems.

Let us start with Zipf's Law. George Kingsley Zipf (1932) was a linguist who noticed a striking regularity. If you order words in a language according to how frequently they are used, the second word is used about half as frequently, the third word about a third as frequently, and so forth. For example, in one well-known compilation of English word frequency, the word "the" accounts for approximately 7% of the words, "of" appears about half that frequently, "and" about a third as frequently, and so on. It was also noted that the same kind of distribution describes phenomena unrelated to linguistics, such as the sizes of cities, personal income, and the sizes of corporations.

Quantities that follow Zipf's distribution can easily be recognized by the straight line produced when one plots the log of the rank by the log of the quantity. Figure 18.3 shows the relation between the log of word rank and the log of word frequency.

Zipf's Law can be generalized to a family of probability distributions, called *power-law* distributions or *Pareto* distributions. Instead of ranking the data from high to low, calculate the proportion of instances of the phenomena that fall into various ranges. Highly ranked objects will fall into the extreme right of the frequency (or probability) distribution.

Pareto distributions are highly skewed and dominated by extreme values. In part it is this domination by their extreme values that has captured the interest of social and other scientists. Much economic modeling before the 'great recession of 2008' contained the assumption that exogenous disturbances were normally distributed. It's well known that in normal distributions most of the observations are close to the mean. Therefore the models were not prepared for the occurrence of very large and unexpected events, like the collapse of the housing bubble. If exogenous events are normally distributed, such large unexpected events will not occur.

Let's explore this idea further by comparing four illustrative distributions: the normal, the log normal, the exponential, and the Pareto.

1. *The normal distribution*: The familiar normal distribution has two parameters, the mean μ and the standard deviation σ. Scores tend

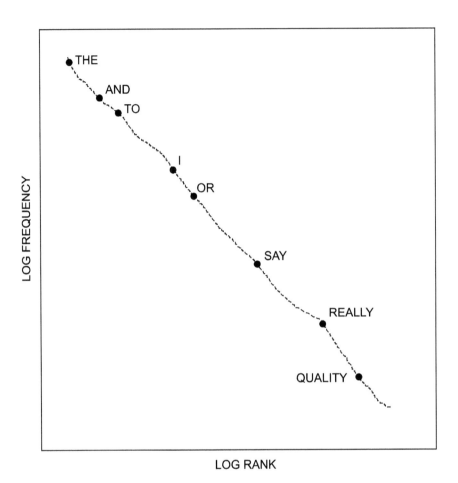

Figure 18.3. Zipf's Law of word frequency.

to be tightly bunched around the mean: approximately 95% of the observations will lie within two standard deviations of the mean.

$$f(x) = \frac{e^{-\frac{(x-\mu)^2}{2\sigma^2}}}{\sqrt{2\pi}\,\sigma} \qquad (18.6)$$

2. *The log-normal distribution*: A variable has the log-normal distribution if the logarithm of its values is normally distributed. The log-normal distribution is skewed, and extremely large values are more likely than for the normal distribution.

$$f(x) = \frac{e^{-\frac{(-\mu+Log[x])^2}{2\sigma^2}}}{\sqrt{2\pi}\,x\sigma} \qquad (18.7)$$

3. *The exponential distribution*: The (negative) exponential distribution is highly skewed and asymmetric with just one parameter α indicating

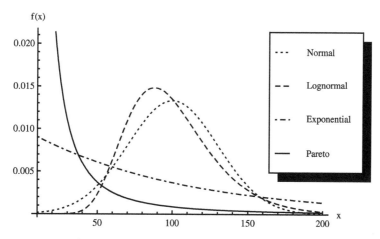

Figure 18.4. Four distributions, with the same mean.

how quickly the values fall off: the greater α, the more quickly the frequencies decline.

$$f(x) = e^{-x\alpha}\alpha \tag{18.8}$$

4. *The Pareto distribution*: Like the exponential distribution, the Pareto distribution is highly skewed, with most of its values small. There are two parameters, α and β. The parameter α is the minimum value of the distribution, and β gives the rapidity with which frequencies decline.

$$f(x) = x^{-1-\beta}\alpha^{\beta}\beta \tag{18.9}$$

Figure 18.4 shows examples of these four distributions. The mean of all four distributions is 100. The standard deviation of the normal and log normal distributions is 30. The mean and standard deviation of the exponential distribution must be equal, and so they both equal 100. The standard deviation of the Pareto distribution is infinity.

All the characteristics of the four distributions that we are going to discuss can be illustrated with the demonstration *Pareto Distribution Comparision*. Comparing the normal and log-normal distributions, we can see that at a little less than 159, values for the log-normal distribution start to become more frequent than those of normally distributed variables: the density is greater, the percentage of the population with a more extreme value is higher. The log-normal distribution has a *fatter* tail in the sense that there is a point beyond which the values in one distribution are more frequent. The exponential distribution also has a fatter tail than the log-normal distribution: at a value slightly more than 157 the frequencies of the exponential distribution are uniformly higher than those of the log-normal distribution.

Finally, at a value slightly less than 636, frequencies for the Pareto distribution become greater than those for those for the exponential distribution. About 1% of the Pareto-distributed values are greater than this, while only one-sixth of 1% of the observations in an exponential distribution are greater.

However, these figures do not do justice to the extreme skewness of the Pareto distribution. Extreme values can magnify the differences between the distributions. That in 2005 the top 1% in the United States received 21.8% of all income reflects the income levels of the super rich, not just the income it takes to reach the 99th percentile. A distribution can describe a resource, and we can ask what percentage of the resource is attributed to the top values. This is what the last column of the *Pareto Distribution Comparison* demonstration does.

$$D(y) = \frac{\int_y^\infty x f(x) dx}{\int_{-\infty}^\infty x f(x) dx} \tag{18.10}$$

For the Pareto distribution this highest 1% with values greater than 160 has 63% of the resources. For the exponential distribution the highest 1% (with values above 460) has less than 6% of the resources. This is the extreme domination compared to the more commonly used normal distribution by observations having large values that have captured the interest of social and physical scientists. The Pareto distribution allows for the possibility of cataclysmic events (falls in the stock market, massive earthquakes, mass biological extinctions in the fossil record) that would be inconceivable if these were normally distributed.

Models have been created to help us understand why complex systems might produce Pareto distributions out of their own inner workings, without any exogenous influences. These models all depend on patterns of interdependence among the many objects in the complex system so that changes can cascade and multiply in an explosive manner. These models are not necessarily realistic descriptions of what they apparently describe; they explore the consequences of interdependence in systems.

One of the best known is a simple model of avalanches, devised by the physicist Per Bak. Consider grains of sand falling one by one onto a flat rectangular surface. Piles of sand build up until a certain level of tension has been reached and then, unpredictably, there is an avalanche, a small one or a large one that has more of a leveling effect. The size of the avalanches may follow a power law, depending on the size and shape of the grains.

A power law of avalanche size can also be generated from a very simple computer simulation, implemented in the demonstration *Sand Pile*. The rules of this simulation are quite simple. At the beginning of each cycle a cell in a rectangular grid is randomly selected, and a counter for the number of "grains" of sand in that cell is increased by one. If that cell totals four grains, it is emptied and its grains distributed to its neighboring cells. If any of those cells now have four grains their grains are redistributed. This process continues until there are no unstable cells. Some of the grains of sand redistributed from cells at the edge of the square disappear "over the edge." The demonstration counts the total sizes of avalanches and the number of grains (if any) that fall off the grid. The user can determine the size of the square and the number of grains of sand dropped. The output includes the time record of avalanche size and losses over the edge, the distribution of sizes of these two events, and their log log plots. The user of this demonstration can also view the entire sequence from beginning to end. To get approximately a Pareto distribution of avalanche size, run at least 2,000 time periods.

Another example of the power law concerns extinctions in the fossil record. The existence of very large simultaneous die-offs of many species is usually explained by exogenous events, like large meteors hitting the earth, causing prolonged periods of cold weather. On the other hand, many models have been suggested that produce large-scale extinction events from the mere fact that an ecological system consists in a complex web of interconnected species, where the extinction of one species can lead to the extinction of species that depend on it, which can lead to further extinctions, and so forth.

One intriguing model was suggested by two physicists, Nunes Amaral and Meyer (1999). They propose that such interdependencies can by themselves lead to mass extinctions. In their model, implemented in *Amaral-Meyer*, there are six levels of species arranged in a hierarchy. Higher level species eat lower level species, with the exception of the lowest level (representing plants). Each species at a higher level is connected to at most three species at the immediately lower level. A fraction of the species at the lowest level is randomly selected for extinction. All species at the second level, all of whose prey becomes extinct, also become extinct, and so on to higher levels. New species are created to replace species that become extinct. When you run this demonstration you will see that the number of extinctions per time period is highly skewed. The numbers roughly follow a power law.

Whether or not there is a common mechanism producing power laws in all complex systems or whether indeed all complex systems have anything in common other than their complexity remains to be seen; the search for such properties may prove to be quixotic. However, the insight that systems can have unexpected emergent properties that can fruitfully be explored with computer simulations is a profoundly valuable insight.

Chapter Demonstrations

- *Classic Logistic Map* shows the values of the parameter μ for which the logistic sequence is chaotic and, for those values for which it not chaotic, what the sequence converges to.
- *Pareto Distribution Comparison* computes the Pareto distribution with two other skewed distributions and the normal distributions.

- *Logistic Sequence Sensitivity* shows how tiny differences in initial conditions can lead to exponentially increasing differences in two sequences.
- *Sand Pile* is a reproduction of the Bak simulation of sandpile avalanche size. The sizes of avalanches follow a Pareto distribution.
- *Amaral-Meyer* is the Amaral-Meyer mode of mass extinctions in evolutionary history brought about by complex patterns of interdependence among species.

EXERCISES

1. Using the demonstration on the sensitivity of the logistic sequence, show that the logistic map converges to just one value when the multiplier is less than 3, to two values when it is between 3 and 3.5, and to four values when it is a little greater than 3.5.
2. Using 2,000 or so time periods, verify that the number of extinctions approximately follows a power-law distribution.
3. Using as a criterion the proportion of resources controlled by the top 5% of the observations, rank order the four distributions in terms of how unequal they are.

AFTERWORD: "RESISTANCE IS FUTILE"

These days, some sociologists may look upon the role of mathematics in their discipline much like Starfleet looked upon the first encroaching Borg cubes. In *Star Trek*, the Borg are a race of cybernetic organisms that "assimilated" any alien race that they stumbled upon. Hapless starship captains would sit on the bridge of their ship while the power went out, red alert lights flashed, and an ominous voice intoned, "We are the Borg, we will add your biological and technological distinctiveness to our own, resistance is futile."

Anyone could look at the Borg race and question whether they really preserved the "biological and technological distinctiveness" of the races they assimilated. For starters, they're all ugly and grey (the exception being Seven of Nine, of course). Secondly, most of their ships are boring cubes. Thirdly, they have no individuals; only the collective exists. What could be more terrifying to a pop culture that celebrates the individual than to lose all uniqueness?

Social life and everything that comes with it is, without doubt, rich and complex. That is why many postmodern sociologists are also terrified with dealing with the math as they know it. Many believe math means quantitative methods, and quantitative methods mean numbers, and numbers destroy all the nuance and color of daily social interaction. For example, how do you capture how people negotiate order in the daily institutions to which they belong? How can a number represent a pattern of action? Can mathematics represent abstract concepts such as social power, masculinity, or exploitation? To some in the field, social phenomena are too rich and fluid to be studied by applying math or even the scientific method. It also doesn't help ease the fears of sociologists when our neighbors in the nearby star system of economics have already been "assimilated" by mathematics.

However, the extent to which mathematics can be used to break down complexity is limited by one's knowledge and maturity in mathematics, just as the extent to which computers can help one's research is limited by one's capability in programming or using software packages. So, mathematics is not quite like the Borg. It's not all boring and grey—we just have to learn how to see color. However, it is like the Borg in one respect: assimilation is imminent and resistance is futile.

A trend to be celebrated, not feared, is that intellectual life in all the sciences is more and more dominated by mathematics and the use of computers to collect and analyze data and simulate models. The more mathematics you know and the greater your facility in the use of computers, the better prepared you will be for the future. The omnipresence of computers has had a profound effect: more and more data are available, but so are tools to analyze complex systems of relationships. Properties of these complex systems of human relations are the defining quintessential topic of sociology. Increasingly there will be an integration of computer science and sociology. There is already a journal devoted both to social networks and data mining.

There are a number of ways in which computers improve sociology. More and more data are being automatically collected and potentially available for analysis as our interactions with one another are increasingly mediated by the computer. Email, Facebook, Twitter, and other computer-mediated forms of interaction generate invaluable traces of interaction patterns. These interaction patterns are almost impossibly large and complex. They require computers for their analysis for scientific, economic, or political purposes. Large retail chains analyze customer records carefully to target their advertisements. The essence of the highly successful Google search algorithm is to use user search patterns to inform future search results. Facebook uses personal information that its millions of users provide to direct advertisements. The government uses network and data-mining algorithms to detect suspicious clusters of communications in phone calls. There is a large and growing market for those who are skilled in analyzing large complex network data sets.

Computer simulations of complex social systems that cannot be resolved by exact mathematical methods are increasingly being analyzed in part through computer simulations. More and more contributions to the leading journals for mathematical sociology (such as *Journal of Mathematical Sociology* or *Social Networks*) routinely use computer simulations to draw inferences from mathematical models. The mathematics necessary to draw implications from many complex models is simply undeveloped.

A good mathematical background is not less important just because computers are playing more of a role in mathematical sociology. The models that are simulated are usually described in mathematical language, and properties of these are examined only when no exact mathematical solution is available. If you've read and enjoyed this book, you may be asking yourself how you can prepare yourself to learn more. The first thing you should do is to take more mathematics courses. In a course on linear algebra, or finite mathematics, you will learn about the properties of matrices. Matrices are an indispensable way of describing networks, Markov chains, and statistics. Probability theory and stochastic processes play an essential role in many mathematical sociology models. Game theory is the mathematics of strategic interaction between rational actors. Although we know that human actors are not rational, rational action can be a useful baseline model for studying social behavior. The study of combinatorics is a part of probability whose usefulness extends far beyond calculating the probabilities of drawing poker hands. For example, the explanation of the binomial distribution in this book depended on an idea from combinatorics: the number of ways K objects can be selected from N objects. To a mathematician, algebra does not mean solving numerical equations with unknowns. It is the study of abstract mathematical structures with a variety of objects and operations. You have been exposed to one algebra, linear algebra, when you studied matrices, but there are many more. There has been exciting work in mathematical sociology using the algebra of groups and semigroups to analyze patterns of relations in networks. Finally, calculus is the basic language of science. Differential equations are the way in which calculus is used to study change in systems. Although they are not as commonly used in sociology as in the physical sciences, a basic course in differential equations can be useful.

Mathematical sociologists have also borrowed heavily from different fields containing models and mathematics that can be modified to have sociological applications. Microeconomic models are based on rational choice, and economists are well trained in game theory. Physicists are becoming active contributors to mathematical sociology, particularly to the study of networks. People are obviously not identical like electrons, but the tools that physicists have developed for handing systems could be of use. Computer scientists study networks, like the Internet, which are complex. The tools they develop can be helpful to sociologists. Linguists have developed mathematical models for language that could prove informative in the study of culture and social structure. The mathematical tools that epidemiologists have developed to study the spread of disease can be helpful for studying the transmission of information, rumors, or culture through social networks. Geographers have developed computer techniques for illustrating and studying special distributions. These tools have been used by sociologists.

There are also journals you could look at to get some sense of the issues that interest mathematically oriented sociologists.

- The *Journal of Mathematical Sociology* publishes the greatest variety of mathematical models in sociology
- *Social Networks* publishes mathematical, methodological, and substantive papers on social networks
- *Rationality and Society* specializes in applications of game theory and rational choice to social life
- *Social Network Analysis and Mining* is a new journal specializing in the analysis of large data sets
- *Science*, published by the American Association for the Advancement of Science and Nature, the leading international interdisciplinary journal of science, not infrequently publish important papers on networks

In 1900 the famous German mathematician David Hilbert presented a collection of important unsolved problems whose attempts of solution shaped the course of 20th-century mathematics. Progress toward the solution to one of Hilbert's problems always leads to fame and prizes. We can't guarantee that solutions to the following problems will lead to international fame and fortune, but many sociologists would be grateful for progress on the following problems.

- Regarding networks and homophily, birds of a feather flock together, but people are influenced by those they like. Both these processes result in the same outcome (similar people together in groups), but there is no standard accepted way of separating these two processes (Chapter 6).
- There are lots of models that show how groups arrive at consensus but no generally accepted model of how groups become polarized or how two groups can become more and more different and possibly hostile (Chapters 6 and 13).
- We know that people are affected by their positions in networks, but we don't have a variety of models of how people create their networks. We also do not have good models for network change and evolution (Chapters 6 and 9).
- What are the most important mechanisms for bringing about co-operation in groups? In particular, how do groups avoid succumbing to the free rider problem in the provision of public goods? Too many answers, many more than are covered in this book, have been proposed for the solution to this problem, and there is no consensus (Chapter 16).
- There are a variety of measures of centrality in networks, but there are no well-established criteria for when one measure is preferable to another (Chapter 10).
- Are human groups unpredictable because humans themselves are complex organisms or because there are truly chaotic dynamics in groups (Chapter 18)?

BIBLIOGRAPHY

R. Albert and A. L. Barabási. Statistical mechanics of complex networks. *Reviews of Modern Physics*, 74(1): 47–97, 2002. ISSN 15390-756.

Peter S. Bearman, James Moody, and Katherine Stovel. Chains of affection: The structure of adolescent romantic and sexual networks. *American Journal of Sociology*, 110(1): 44–91, 2004. ISSN 00029602. http://www.jstor.org/stable/10.1086/386272.

E. A. Bender and E. R. Canfield. The asymptotic number of labeled graphs with given degree sequences. *Journal of Combinatorial Theory, Series A*, 24(3): 296–307, 1978. ISSN 0097-3165.

A. Binder. Constructing racial rhetoric: Media depictions of harm in heavy metal and rap music. *American Sociological Review*, 58(6): 753–767, 1993. ISSN 0003-1224.

P. Bonacich and G. W. Domhoff. Latent classes and group membership. *Social Networks*, 3(3): 175–196, 1981.

R. S. Burt. *Structural holes: The social structure of competition*. Harvard University Press, 1995. ISBN 0674843711.

D. Cartwright and F. Harary. Structural balance: A generalization of Heider's theory. *Psychological Review*, 63(5): 277–293, 1956. ISSN 0033-295X.

J. S. Coleman, E. Katz, H. Menzel, and Columbia University, Bureau of Applied Social Research. *Medical innovation: A diffusion study*. Bobbs-Merrill, 1966.

J. A. Davis. Clustering and hierarchy in interpersonal relations: Testing two graph theoretical models on 742 sociomatrices. *American Sociological Review*, 35(5): 843–851, 1970. ISSN 0003-1224.

P. S. Dodds, R. Muhamad, and D. J. Watts. An experimental study of search in global social networks. *Science*, 301(5634): 827–829, 2003.

R. M. Emerson. Social exchange theory. *Annual Review of Sociology*, 2:335–362, 1976. ISSN 0360-0572.

A. Giddens. *The consequences of modernity*. Polity Press, 1990.

M. S. Granovetter. The strength of weak ties. *American Journal of Sociology*, 78(6): 1360–1380, 1973. ISSN 0002-9602.

F. Harary. On the measurement of structural balance. *Behavioral Science*, 4(4): 316–323, 1959. ISSN 1099-1743.

F. Heider. Attitudes and cognitive organization. *Journal of Psychology*, 21(1): 107–112, 1946. ISSN 0022-3980.

P. G. Hoel, S. C. Port, and C. J. Stone. *Introduction to stochastic processes*, Houghton Mifflin, 1972.

G. C. Homans. Social behavior as exchange. *American Journal of Sociology*, 63(6): 597–606, 1958. ISSN 0002-9602.

J. G. Kemeny, J. L. Snell, and G. L. Thompson. *Introduction to finite mathematics*. PrenticeHall, 1966. ISBN 0134837843.

R. Kumar, P. Raghavan, S. Rajagopalan, D. Sivakumar, A. Tomkins, and E. Upfal. Stochastic models for the web graph. *Proceedings of the 41st Annual Symposium on Foundations of Computer Science (FOCS) Redondo Beach*, 57–65, 2000.

P. H. Leslie. On the use of matrices in certain population mathematics. *Biometrika*, 183–212, 1945. ISSN 0006-3444.

R. D. Luce and H. Raiffa. *Games and decisions: Introduction and critical survey.* Dover, 1989. ISBN 0486659437.

W. Nooy, A. Mrvar, and V. Batagelj. *Exploratory social network analysis with Pajek.* Cambridge University Press, 2005.

Luís A. Nunes Amaral and Martin Meyer. Environmental changes, coextinction, and patterns in the fossil record. *Physical Review Letters*, 82(3): 652–655, 1999. doi:10.1103/PhysRevLett.82.652.

J. F. Padgett and C. K. Ansell. Robust action and the rise of the Medici, 1400–1434. *American Journal of Sociology*, 98(6): 1259–1319, 1993. ISSN 0002-9602.

M. Puzo. *The Godfather.* Fawcett, 1970.

F. J. Roethlisberger, W. J. Dickson, and H. A. Wright. *Management and the worker: An account of a research program conducted by the Western Electric Company, Hawthorne Works, Chicago.* Harvard University Press, 1939. ISBN 0674546768.

W. G. Roy and P. Bonacich. Interlocking directorates and communities of interest among American railroad companies, 1905. *American Sociological Review*, 53 (3):368–379, 1988. ISSN 0003-1224.

T. C. Schelling.Models of segregation. *American Economic Review*, 59(2):488–493, 1969. ISSN 0002-8282.

Thomas Schelling. Models of segregation. *Journal of Mathematical Sociology*, 1(1):143–86, 1972.

J. Travers and S. Milgram. An experimental study of the small world problem. *Sociometry*, 32(4):425–443, 1969. ISSN 0038-0431.

T. Veblen. *The theory of the leisure class. 1899.* Forgotten Books, 1965. ISBN 1606801805.

S. Wasserman and K. Faust. *Social network analysis: Methods and applications.* Cambridge University Press, 1994. ISBN 0521387078.

D. J. Watts and S. H. Strogatz. Collective dynamics of small-world networks. *Nature*, 393 (6684):440–442, 1998. ISSN 0028-0836.

G. Zipf. *Selective studies and the principle of relative frequency in language.* Harvard University Press, 1932.

INDEX

adjacency matrix, 69
age-specific, 171
antisymmetric, 42
arcs, 53
arrays, 67
associativity, 18
attack, 131
avalanche, 210
Axelrod, 184

balance theory, 137
bank wiring room, 91
Barabasi-Albert model, 121
bell-shaped, 118
betweenness centrality, 96
bi-directional, 58
binding agreements, 193
binomial coefficient, 32
binomial distribution, 32
bipartite, 58
Boole, 18
Boolean algebra, 18
Boolean lattice, 18
bridge, 61
Burt, 49
butterfly effect, 202

Cartesian product, 38
cascade, 133
centrality, 90
centralization, 99
chaos, 202
characteristic value, 193
classification, 16
clique, 85
cliques, 21, 79
closeness centrality, 94
clustering, 110
clustering coefficient, 108
coalitions, 196
cohesion, 21, 53
combination, 31
community structure, 87
commutative, 75
commutivity, 18
complement, 15

complete network, 55
complexity, 206
component, 58, 80
conditional probability, 29
configuration model, 127
conspicuous consumption, 118
containment, 13
convergence, 3
cooperate, 180
copy model, 125
core, 193, 199
critical value, 135
cross-classification, 16
cut points, 98
cycle, 56, 139

defection, 181
degree, 53
degree centrality, 93
demographic transition, 162
demography, 161
density, 5, 55, 106
deterministic, 203
deterministic chaos, 206
deterministic non-chaos, 205
diagram, 40
disjoint, 14, 17
distributivity, 18
dyadic relation, 38

edges, 53
ego network, 108
eigenvector centrality, 95
element, 13
elementary events, 25
emergent property, 6
empty, 14
epidemiological model, 135
equilibrium outcome, 205
equilibrium vector, 157
equivalence relation, 46
events, 25
evolutionarily stable, 185
evolutionary tournaments, 185
evolving ecology, 185
excess, 195

exchange network, 192
expected value, 28
exponential distribution, 208

failure, 131
fair, 29
farness, 94
fertility rate, 162
flow, 97
forbidden triad, 64

geodesics, 56
Giddens, 115
Granovetter, 63
graph center, 94
graph theory, 53
graphs, 53
groups, 84

Heider, 137
hierarchical diagram, 18
homomorphism, 20
hubs, 132

idempotency, 18
image lattice, 20
in-degree, 59
independence, 30
influence, 93
interlocking directorates,
    12, 19
intermediate state, 155
intersection, 14
iterated prisoner's dilemma, 184

K-plex, 86
kernel, 193, 195
Kevin Bacon, 105

Leslie matrix, 171
life expectancy, 167
local bridge, 62
log-normal, 121
log-normal distribution, 208
logic, 22
logistic curve, 3
loops, 57
Lorenz, 202

main diagonal, 68
Malthus, 161
MAN labeling, 146
mapping, 20
Markov, 149
Markov assumption, 156
Markov chain, 149

maternity rate, 173
mathematical sociology, 1
mathematical statement, 22
matrix, 67
matrix product, 74
Medici, 91
membership, 19
micro, 144
Milgram experiment, 103
mixed strategy, 183

N-clique, 86
Nash equilibrium, 183
negative relationship, 139
network damage, 129
networks, 53
nodes, 53
non-cooperative game, 183
normal distribution, 207

odds, 25
ordered pair, 38
out-degree, 59

Pareto, 118
Pareto distribution, 209
path, 55
payoff, 181
permutation, 31
popularity, 53, 93
population equilibrium, 178
population projection, 173
power-dependence principle, 196
power-dependence theory, 195
power-law distribution, 117, 207
power set, 14, 19, 25
preferential attachment, 122
prisoner dilemna, 180
probability, 25
probability matrix, 71
public good, 183
punishment, 181
pure strategy, 183

radix, 171
random network, 2
random walk, 203
rational actor, 193
reflexivity, 43
relations, 39
replacement, 29
reproduction number, 135
residential segregation, 6
reward, 181
rewiring, 113

scalar, 67
Schelling, 6
sets, 12
simple network, 57
skewness, 210
small-world phenomenon, 103
social capital, 90
stability, 156
stochastic, 203
stratification, 118
strong ties, 64
structural balance, 145
structural configuration, 137
structural equivalence, 48
structural hole, 49
subset, 13
sucker payment, 181
susceptible, 135
symmetric, 41

tail, 209
temptation, 181
tit-for-tat, 185
transition matrices, 149

transitive closure, 49
transitivity, 44, 65
transpose, 72
traverse, 55
triad census, 146
triads, 145
truth, 21
Type I error, 34
Type II error, 34

union, 14
universal set, 15

vector, 67
Venn diagram, 15
vertex, 53

walk, 55
Watts and Strogatz model, 111
weak order, 45
weak ties, 64

Zipf, 118
Zipf's law, 207

Milton Keynes UK
Ingram Content Group UK Ltd.
UKHW052217220924
448588UK00002B/23

9 780691 145495